State-of-the-Art CMOS and MEMS Devices

State-of-the-Art CMOS and MEMS Devices

Editor

Zhiming Chen

MDPI

Basel • Beijing • Wuhan • Barcelona • Belgrade • Novi Sad • Cluj • Manchester

Editor
Zhiming Chen
School of Integrated Circuits
and Electronics
Beijing Institute of
Technology
Beijing
China

Editorial Office
MDPI
St. Alban-Anlage 66
4052 Basel, Switzerland

This is a reprint of articles from the Special Issue published online in the open access journal *Micromachines* (ISSN 2072-666X) (available at: www.mdpi.com/journal/micromachines/special_issues/SotA_CMOS_MEMS).

For citation purposes, cite each article independently as indicated on the article page online and as indicated below:

Lastname, A.A.; Lastname, B.B. Article Title. *Journal Name* **Year**, *Volume Number*, Page Range.

ISBN 978-3-7258-0842-7 (Hbk)
ISBN 978-3-7258-0841-0 (PDF)
doi.org/10.3390/books978-3-7258-0841-0

Contents

About the Editor

Zhiming Chen

Prof. Zhiming Chen received his B.Eng. degree in Electronic Science and Technology from Tsinghua University, Beijing, China, in 2007, and his Ph.D. in Electrical and Computer Engineering from the University of California at Irvine, Irvine, USA, in 2012. In the spring of 2012, he joined as a faculty member at the Beijing Institute of Technology, China, where he is currently a Professor and Associate Dean for the School of Integrated Circuits and Electronics. He also serves as the Director of the Beijing Engineering Research Center of Silicon-based High-Speed System-on-Chip. His research interests include analog, RF, and millimeter-wave (MMW) integrated circuit designs, advanced packaging, and 3D integrated circuits.

micromachines

Editorial

Editorial for the Special Issue on State-of-the-Art CMOS and MEMS Devices

Zhiming Chen

School of Integrated Circuits and Electronics, Beijing Institute of Technology, Beijing 100081, China; czm@bit.edu.cn

1. Introduction

Complementary Metal Oxide Semiconductor (CMOS) and Micro-Electro-Mechanical System (MEMS) devices play significant roles in emerging research fields such as artificial intelligence (AI) [1], 5G communication [2,3], the Internet of Things (IoT) [4], and satellite communication [5,6]. With the continuous development of material science and micro-fabrication processes, high-performance CMOS and MEMS devices have been the subject of extensive investigations and have advanced significantly [7,8]. In recent years, the rapid development of heterogeneous integration [9] and complex multi-functional systems has resulted in higher requirements for CMOS and MEMS devices; it is thus urgent to develop novel CMOS integrated circuits (ICs) and MEMS devices utilizing advanced materials, techniques, and processes. Thus, the research in this field is currently focused on the state-of-the-art design [10–12], modeling [13], fabrication [14], and application of CMOS and MEMS devices [15,16].

This Special Issue, entitled "State-of-the-Art CMOS and MEMS Devices", presents crucial research progress in the area of advanced CMOS ICs and MEMS devices. The Special Issue contains 10 published articles (9 research articles and 1 review article) that address the urgent demand for high-performance silicon ICs and MEMS devices; these articles aim to further advance the field through extensive explorations of device and circuit design techniques, performance optimization methods, and system integration strategies.

2. Overview of the Published Articles

The development of software-defined radio, 5G mobile communication, satellite communication, and wireless sensor networks has augmented the demands for high-performance CMOS radio frequency (RF) transceiver front-ends and their key building blocks, such as low-noise amplifiers (LNAs), power amplifiers (PAs), and circulators. To enable the utilization of multi-standard wireless communication systems that support several applications simultaneously, Wan et al. (contribution 1) proposed the application of an inductorless and gain-controllable 0.5~2.5 GHz wideband LNA based on second-generation current-controlled current conveyors (CCCIIs). By varying the DC biasing current of the CCCII, the voltage gain of the proposed LNA is controllable in the range of 1~18 dB. The design achieves a typical voltage gain of 12.6 dB with a gain ripple of $\pm$1.5 dB, a noise figure of 4.05~4.35 dB across the whole band of 0.5~2.5 GHz, and an input third-order intercept point (IIP3) of -2.5 dBm at 1.5 GHz. By utilizing a 0.18 μm RF CMOS, the LNA occupies an active chip area of 0.096 mm^2, with a power consumption of 12.0 mW.

Because the specification of PAs is closely related to variations in the humidity, Zhou et al. (contribution 2) conducted an experimental study of PA specifications under different humidity conditions to illustrate the relationship between the degradation of PA performance and humidity conditions. The paper presents results regarding the degradation of a PA subjected to different humidity levels, where the S_{21} and output power decrease with the increase in humidity. Furthermore, it also analyzes the main reason for this degradation, which is the decrease in the oxide capacitance and increase in the threshold voltage as the

Citation: Chen, Z. Editorial for the Special Issue on State-of-the-Art CMOS and MEMS Devices. *Micromachines* **2024**, *15*, 327. https://doi.org/10.3390/mi15030327

Received: 22 February 2024
Accepted: 23 February 2024
Published: 27 February 2024

humidity increases, resulting in a reduction in the transconductance and an increase in the on-resistance.

For those who are seeking a broad overview and a general understanding of the design of PAs in CMOS technology, the review article by Hamid et al. (contribution 3) could be a good reference. The authors provide a state-of-the-art review on CMOS RF PAs for wireless communication systems. RF PAs in a CMOS transmitter play a crucial role in amplifying RF signals and transmitting them through the antenna. This review paper presents a concise discussion of the performance metrics, which is followed by an overview of the research trends observed in CMOS PA design techniques with a focus on efficiency, linearity, and bandwidth enhancement. Although the on-chip transformer and out-phasing techniques are capable of delivering the watt-level maximum output power for CMOS PA, they are still limited in delivering the watt-level linear output power.

Full-duplex (FD) communication enables simultaneous transmission and reception; however, the implementation of on-chip non-magnetic passive circulators is a great challenge. Gao et al. (contribution 4) designed and implemented a non-magnetic on-chip passive circulator operating at the Ku band in a 90 nm bulk CMOS using switch-based spatio-temporal conductivity modulation. Owing to the four-phase non-overlapping clock signal, the proposed circulator achieves a 3.9 dB transmitter (TX)-to-antenna (ANT) and a 4.0 dB ANT-to-receiver (RX) insertion loss. The TX-to-RX isolation is better than 17.2 dB, and the TX-to-ANT IIP3 and ANT-to-RX IIP3 values are 19.7 dBm and 20.0 dBm, respectively, while occupying a die area of 1.55 mm $\times$ 1.15 mm. To further develop CMOS ICs in a millimeter-wave range, He et al. (contribution 5) present a D-band direct conversion IQ receiver with an on-chip multiplier chain. The D-band LNA is implemented using a gain-boosting and stagger-tuning technique to provide high gain and a large bandwidth. The X9 multiplier chain including Marchand balun and a quadrature (90°) hybrid is employed to provide four LO signals to drive the IQ mixer. This receiver is implemented in a 130 nm SiGe process and consumes a core area of 1.04 mm^2. The proposed receiver exhibits a 20 GHz bandwidth from 150 GHz to 170 GHz, with a conversion gain of 28 dB and NF of 7.3 dB at 158 GHz.

Yang et al. (contribution 6) present an ultra-low-power analog multiplier–divider that is compatible with digital code words and is applicable to the integrated structure of resistive random-access memory (RRAM)-based computing-in-memory (CIM) macros. The multiplication and division of the current are accomplished by a current-mirror-based structure, improving the energy efficiency. Designed and fabricated in a 55 nm CMOS process, the proposed work achieves 8-bit precision for analog current multiplication and division operations. The measurement results show that the signal delay is 1 μs when performing the 8-bit operation. The power consumption is less than 6.15 μW with a 1.2 V supply voltage.

The increasing complexity of the power supply network has led to the susceptibility of the power supply port to electromagnetic interference (EMI). In order to obtain a deep understanding of the electromagnetic susceptibility (EMS) mechanism and guide the design of the electromagnetic compatibility (EMC) of operational amplifiers (op-amps), Huang et al. (contribution 7) assessed the effect of EMI on the performance of op-amps through the power supply port by employing a bulk current injection (BCI) method. The authors proposed a novel method of conducted susceptibility and obtained the susceptibility threshold regularities of the op-amps at the power supply port under the interference, which provided evidence that EMI reduced the reliability of the op-amp by affecting the offset voltage of op-amps; this also demonstrated that the sensitivity type of op-amps is peak sensitive at the power supply port.

Hot-carrier degradation (HCD) has been recognized as the most harmful degradation issue limiting the lifetime of modern metal oxide semiconductor field-effect transistors (MOSFETs). Tyaginov et al. (contribution 8) developed a compact physical model for HCD that is valid over a wide range of gate–source (V_{gs}) and drain–source voltages (V_{ds}) based on the refined modeling of the carrier transport for both primary and secondary carriers.

Special attention is paid to the contribution of secondary carriers (generated by impact ionization) to HCD, which is shown to be significant under stress conditions with low V_{gs} and relatively high V_{ds}. To validate the model, foundry quality n-channel transistors were employed, with a broad range of stress voltages.

To facilitate the integration of RF ICs and MEMS devices and minimize the form factor of electronic systems, Dong et al. (contribution 9) propose novel integrated passive devices (IPDs) that include capacitors, inductors, and a bandpass filter designed by employing through-silicon via (TSV) structures. The influence of the structural parameters of TSVs on the electrical performance of the TSV-based capacitor and inductor are evaluated. Moreover, a compact third-order Butterworth bandpass filter with a central frequency of 2.4 GHz is developed, and the footprint is only 0.814 mm $\times$ 0.444 mm. The simulated 3 dB bandwidth of the filter is 410 MHz, and the fraction bandwidth (FBW) is 17%. In addition, the in-band insertion loss is less than 2.63 dB, and the return loss in the passband is better than 11.4 dB, thus exhibiting good RF performance.

With the development of space laser communication and satellite internet constellations, the demand for microminiature laser communication terminals is expanding. To meet the requirements of size, weight and power (SWaP), miniaturized terminals require smaller drive components in order to complete on-orbit scanning and capture. Wang et al. (contribution 10) propose a laser scanning capture model based on a MEMS micromirror that can provide rapid, large-scale scanning analysis. A scanning overlap factor is introduced to enhance the likelihood of capture under the influence of microvibrations. An experimental analysis validated the effectiveness of the scanning capture strategy, thus establishing a theoretical basis for future ultra-long-distance microspace laser communication.

3. Conclusions

The articles published in this Special Issue present important advancements in the field of CMOS and MEMS devices. I would like to express my sincere gratitude to all the authors, who provided insights and shared their solutions, and I also would like to thank the editors and reviewers who helped to improve the papers and made important contributions to this Special Issue. However, several challenges and obstacles remain to be addressed in the future, such as ultra-wideband CMOS transceivers, CMOS-MEMS integration technology, and high-density multi-chiplet 2.5D/3D-integrated microsystems, etc. We hope that the articles showcased in this Special Issue are interesting and inspiring for its readers, especially young scholars who are eager to learn about recent advances and contribute future research in the field.

List of Contributions

1. Wan, Q.; Liu, J.; Chen, S. An Inductorless Gain-Controllable Wideband LNA Based on CCCIIs. *Micromachines* **2022**, *13*, 1832. https://doi.org/10.3390/mi13111832.
2. Zhou, S.; Yang, C.; Wang, J. Experimental Investigation of Relationship between Humidity Conditions and Degradation of Key Specifications of 0.1–1.2 GHz PA in 0.18 μm CMOS. *Micromachines* **2022**, *13*, 1162. https://doi.org/10.3390/mi13081162.
3. Hamid, S.S.; Mariappan, S.; Rajendran, J.; Rawat, A.S.; Rhaffor, N.A.; Kumar, N.; Nathan, A.; Yarman, B.S. A State-of-the-Art Review on CMOS Radio Frequency Power Amplifiers for Wireless Communication Systems. *Micromachines* **2023**, *14*, 1551. https://doi.org/10.3390/mi14081551.
4. Gao, J.; Wang, X.; Han, F.; Wan, J.; Gu, W. Analysis and Design of a Non-Magnetic Bulk CMOS Passive Circulator Using 25% Duty-Cycle Clock. *Micromachines* **2023**, *14*, 33. https://doi.org/10.3390/mi14010033.
5. He, F.; Ding, Y.; Xu, Z.; Ni, M.; Tian, Y.; Zhang, Z.; Shi, Z.; Wang, K.; Xie, Q.; Wang, Z. A D-Band Direct-Conversion IQ Receiver with 28 dB CG and 7.3 dB NF in 130 nm SiGe Process. *Micromachines* **2023**, *14*, 87. https://doi.org/10.3390/mi14010087.
6. Yang, Y.; Lv, S.; Li, X.; Wang, X.; Wang, Q.; Yuan, Y.; Liang, S.; Zhang, F. An Ultra-Low-Power Analog Multiplier–Divider Compatible with Digital Code for RRAM-Based

Computing-in-Memory Macros. *Micromachines* **2023**, *14*, 1482. https://doi.org/10.3390/mi14071482.
7. Huang, P.; Li, B.; Wei, M.; Hao, X.; Chen, X.; Huang, X.; Huang, W.; Zhou, S.; Wen, X.; Xie, S.; et al. Electromagnetic Susceptibility Analysis of the Operational Amplifier to Conducted EMI Injected through the Power Supply Port. *Micromachines* **2024**, *15*, 121. https://doi.org/10.3390/mi15010121.
8. Tyaginov, S.; Bury, E.; Grill, A.; Yu, Z.; Makarov, A.; Keersgieter, A.D.; Vexler, M.; Vandemaele, M.; Wang, R.; Spessot, A.; et al. Compact Physics Hot-Carrier Degradation Model Valid over a Wide Bias Range. *Micromachines* **2023**, *14*, 2018. https://doi.org/10.3390/mi14112018.
9. Dong, H.; Ding, Y.; Wang, H.; Pan, X.; Zhou, M.; Zhang, Z. Design of a Novel Compact Bandpass Filter Based on Low-Cost through-Silicon-Via Technology. *Micromachines* **2023**, *14*, 1251. https://doi.org/10.3390/mi14061251.
10. Wang, X.; Han, J.; Wang, C.; Xie, M.; Liu, P.; Cao, Y.; Jing, F.; Wang, F.; Su, Y.; Meng, X. Beam Scanning and Capture of Micro Laser Communication Terminal Based on MEMS Micromirrors. *Micromachines* **2023**, *14*, 1317. https://doi.org/10.3390/mi14071317.

Funding: This work was funded in part by the National Natural Science Foundation of China under Grant 92373105, 62074015, and 62101038.

Conflicts of Interest: The authors declare no conflicts of interest.

References

1. Huo, Q.; Yang, Y.; Wang, Y.; Lei, D.; Fu, X.; Ren, Q.; Xu, X.; Luo, Q.; Xing, G.; Chen, C.; et al. A computing-in-memory macro based on three-dimensional resistive random-access memory. *Nat. Electron.* **2022**, *5*, 469–477. [CrossRef]
2. Sadhu, B.; Tousi, Y.; Hallin, J.; Sahl, S.; Reynolds, S.K.; Renstrom, O.; Sjogren, K.; Haapalahti, O.; Mazor, N.; Bokinge, B.; et al. A 28-GHz 32-element TRX phased-array IC with concurrent dual-polarized operation and orthogonal phase and gain control for 5G communications. *IEEE J. Solid-State Circuits* **2017**, *52*, 3373–3391. [CrossRef]
3. Naghavi, S.; Gao, J.; Stadius, K.; Ryynänen, J. A Compact and Wideband mmWave Passive CMOS Circulator Based on Switched All-Pass Networks. *IEEE Microw. Wireless Technol. Lett.* **2024**, *34*, 41–44. [CrossRef]
4. Ghosh, S.; Goh, D.; Koh, Y.; Sharma, J.; Toh, W.; Chen, W.; Zhang, Y.; Ng, E.; Lal, A.; Lee, J. Wake-Up IoT Wireless Sensing Node Based on a Low-G Threshold mems Inertial Switch with Reliable Contacts. In Proceedings of the IEEE 36th International Conference on Miro Electro Mechanical Systems (MEMS 2023), Munich, Germany, 15–19 January 2023; pp. 491–494. [CrossRef]
5. Gültepe, G.; Kanar, T.; Zihir, S.; Rebeiz, G.M. A 1024-Element Ku-Band SATCOM Phased-Array Transmitter with 45-dBW Single-Polarization EIRP. IEEE Trans. Microw. *Theory Tech.* **2021**, *69*, 4157–4168. [CrossRef]
6. Wang, Y.; You, D.; Fu, X.; Nakamura, T.; Fadila, A.; Someya, T.; Kawaguchi, A.; Qiu, J.; Pang, J.; Yanagisawa, K.; et al. A Ka-Band SATCOM Transceiver in 65-nm CMOS with High-Linearity TX and Dual-Channel Wide-Dynamic-Range RX for Terrestrial Terminal. *IEEE J. Solid-State Circuits* **2022**, *57*, 356–370. [CrossRef]
7. Sun, Y.; Li, M.; Ding, Y.; Wang, H.; Wang, H.; Chen, Z.; Xie, D. Programmable van-der-Waals heterostructure-enabled optoelectronic synaptic floating-gate transistors with ultra-low energy consumption. *InfoMat* **2022**, *4*, 10. [CrossRef]
8. Yang, B.; Ding, Y.; Cheng, Z.; Ren, A.; Xiao, L.; Yan, Y.; Zhang, Z.; Chen, Z. Modeling and characterization of annealing-Induced Cu protrusion of TSVs with polyimide liner considering diffusion creep behavior. *IEEE Trans. Electron Devices* **2022**, *99*, 695–701. [CrossRef]
9. Wang, Z. Microsystems Using Three-Dimensional Integration and TSV Technologies: Fundamentals and Applications. *Microelectron. Eng.* **2019**, *210*, 35–64. [CrossRef]
10. Nagulu, A.; Zhuang, Y.; Yuan, M.; Garikapati, S.; Krishnaswamy, H. A Third-Order Quasi-Elliptic N-Path Filter With Enhanced Linearity through Clock Boosting. *IEEE J. Solid-State Circuits* **2023**, *58*, 3351–3363. [CrossRef]
11. Sun, S.; Deng, W.; Jia, H.; Wu, R.; Li, C.; Wang, Z.; Chen, Z.; Chi, B. A Wide Tuning Range Dual-Core Quad-Mode Orthogonal-Coupled VCO with Concurrently Dual-Output Using Parallel 8-Shaped Resonator. *IEEE Trans. Microw. Theory Tech.* **2023**, *71*, 2040–2054. [CrossRef]
12. Wan, J.; Li, X.; Fei, Z.; Han, F.; Li, X.; Wang, X.; Chen, Z. 30-GHz Low-Phase-Noise Scalable Multicore Class-F Voltage-Controlled Oscillators Using Coupled-Line-Based Synchronization Topology. *IEEE Microw. Wirel. Compon. Lett.* **2022**, *32*, 1183–1186. [CrossRef]
13. Shan, Y.; Qian, L.; Wang, J.; Wang, K.; Zhou, P.; Li, W.; Shen, W. Driving Principle and Stability Analysis of Vertical Comb-Drive Actuator for Scanning Micromirrors. *Micromachines* **2024**, *15*, 226. [CrossRef] [PubMed]
14. Chen, X.; Chen, Z.; Xiao, L.; Hao, Y.; Wang, H.; Ding, Y.; Zhang, Z. Fabrication and Electrical Characterization of High Aspect Ratio through-Silicon Vias with Polyimide Liner for 3D Integration. *Micromachines* **2022**, *13*, 1147. [CrossRef] [PubMed]

15. Yang, H.; Ren, A.; Ding, Y.; Xiao, L.; Pan, T.; Yan, Y.; Jiao, W.; Xie, H. A robust lateral shift free (LSF) electrothermal micromirror with flexible multimorph beams. *Microsyst. Nanoeng.* **2023**, *9*, 108. [CrossRef]
16. Ren, A.; Ding, T.; Yang, H.; Pan, T.; Zhang, Y.; Xie, H. An electrothermal micromirror array integrated with thermal convection-based mirror position sensors. In Proceedings of the IEEE International Conference on Micro Electro Mechanical Systems (MEMS 2024), Austin, TX, USA, 21–25 January 2024.

micromachines

MDPI

Article

An Inductorless Gain-Controllable Wideband LNA Based on CCCIIs

Qiuzhen Wan *, Jiong Liu and Simiao Chen

College of Information Science and Engineering, Hunan Normal University, Changsha 410081, China
* Correspondence: wanqiuzhen@hunnu.edu.cn

Abstract: In this paper, an inductorless and gain-controllable 0.5~2.5 GHz wideband low noise amplifier (LNA) based on second generation current controlled current conveyors (CCCIIs) is presented. The proposed wideband LNA utilizes CCCIIs as building blocks to implement the amplifier stage and impedance matching stage. By varying the DC biasing current of the CCCII, the voltage gain of the proposed LNA is controllable in the range of 1~18 dB. In the frequency range of 0.5~2.5 GHz, the post-layout simulation results show that the proposed LNA has a typical voltage gain S21 of 12.6 dB with a gain ripple of ±1.5 dB, an input and output return loss (S11 and S22) of, respectively, −21.4 dB to −16.6 dB and −18.6 dB to −10.6 dB, and a high reverse isolation S12 of −65.2 dB to −39.5 dB. A noise figure of 4.05~4.35 dB is obtained across the whole band, and the input third-order intercept point (IIP3) is −2.5 dBm at 1.5 GHz. Using a 0.18 μm RF CMOS technology, the LNA occupies an active chip area of only 0.096 mm^2 with a power consumption of 12.0 mW.

Keywords: low noise amplifier (LNA); wideband; inductorless; gain-controllable; current conveyor; impedance matching circuit

Citation: Wan, Q.; Liu, J.; Chen, S. An Inductorless Gain-Controllable Wideband LNA Based on CCCIIs. *Micromachines* **2022**, *13*, 1832. https://doi.org/10.3390/mi13111832

Academic Editor: Zhiming Chen

Received: 28 September 2022
Accepted: 22 October 2022
Published: 26 October 2022

1. Introduction

In the last few years, in the software-defined and multi-standard wireless communication systems, a radio frequency (RF) transceiver must be able to support several applications simultaneously, such as 3G/4G/5G mobile communication network, wireless local area network (WLAN), wireless sensor network (WSN), ZigBee (IEEE 802.15.4), and many others [1–4]. Therefore, the research of the wideband RF transceiver front-end is significant and requires the research of RF circuits, whose impedances are matched over wide bands and whose frequency characteristics are stable over several GHz.

As the first active block after the antenna, the wideband LNA constitutes one of the essential elements in a wideband RF front-end [5,6]. The most important parameters of wideband LNA are voltage gain, noise figure (NF), bandwidth, linearity, impedance matching, power consumption, and chip area. The goals in wideband LNA design include providing moderate to high gain with good linearity, minimizing its noise figure to enhance the sensitivity of a receiver, and establishing wideband input/output impedance matching to reduce the return loss. The additional constraints of low power consumption and small chip area are imposed in portable systems.

In the published literature, various wideband LNA circuit techniques in submicron technologies have been proposed, such as a filter matching network amplifier, distributed amplifier, feed-forward noise canceling amplifier, and resistive shunt feedback amplifier. The filter matching network amplifier can consume a small amount of power and achieve wideband input impedance matching. However, the utilization of the filter at the input needs a great deal of inductor components [7–9]. The distributed amplifier is usually used to improve voltage gain at high frequency and therefore can extend the bandwidth, which has the inherent advantages of temperature-insensitive wideband input/output impedance matching and wideband voltage gain, but it dissipates a large chip area owing to the cascade of several stages [10,11]. The feed-forward noise canceling amplifier has been proposed to break the trade-off between input impedance matching and noise figure. However, its bandwidth is limited without the use of the inductor components [12,13]. The resistive shunt feedback amplifier can improve gain flatness and provide wideband input impedance matching, but it is difficult to satisfy noise figure and voltage gain requirements simultaneously [14,15].

As can be seen from the above techniques, in the design of wideband LNAs, the inductor components are extensively used to design impedance matching circuits, resonator networks and to widen the −3 dB bandwidth. However, the on-chip inductor components are large in size and cannot be converted easily from one technology to the next. The on-chip inductor components can result in a significant increase in the chip area and therefore the implementation cost. On the other hand, the feasibility of a wideband LNA to multi-mode and multi-standard is not only dependent on its bandwidth, it should also be able to support variable signal gain according to necessity. For example, when the linearity requirement is more important, a moderate gain is required because the linearity is usually inversely proportional to the gain. When the receiver's NF is required for good performance, the LNA needs to provide high gain to decrease noise contributions of the latter stages. The wideband LNA should be gain-controllable so that it can moderate gain amplification of the strong signals and high gain amplification of the weak signals.

In this paper, a new design for the inductorless gain-controllable wideband LNA using second generation current controlled current conveyors (CCCIIs) has been presented and simulated in a 0.18 μm RF CMOS technology. For the realization of this LNA, three different CCCIIs, as the main building blocks, which are symbolized CCCII 1 to CCCII 3, can provide voltage signal amplification and wideband output impedance matching. Compared to other wideband LNA designs, this kind of LNA can exhibit a smaller chip area, a wider gain-controllable range, and a good high frequency performance.

This paper is organized as follows. In Section 2, the operation principle and some RF applications of current conveyor are discussed, then, the design schematic of a new LNA based on CCCIIs is analyzed. In Section 3, the post-layout simulation results of the proposed wideband LNA are introduced, meanwhile, we present comparisons with the recent works to show the advantages of the new LNA. Finally, a conclusion on this LNA is given in Section 4.

2. Operational Principle and Circuit Implementation

2.1. Current Conveyor Principle of Operation

The current conveyors (CCs) can be used with the other active/passive components in a specific circuit configuration to implement many analogue signal functions [16,17]. Usually, CCs simplify analogue circuit design to implement functions, such as impedance converters, filters, gyrators, inductances, oscillators, and operational amplifiers, etc. In the last few years, CCs have also received much attention for their use as RF basic building blocks [18,19]. The advantages of using CCs in RF circuit design are that they do not require inductor components to make the impedance matching circuit and that the voltage gain can be made controllable by only varying a DC biasing current.

In this paper, the use of second-generation current conveyor (CCII) to implement the wideband RF LNA has been proposed. The CCII is an active device comprised of the three ports, denoted port X, port Y, and port Z. Far from being ideal in their operation, the three ports of the CCII are beset by many parasitic elements, represented by the impedances Z_X (R_X in series with L_X), Z_Y (R_Y in series with C_Y), and Z_Z (R_Z in series with C_Z) at ports X, Y, and Z, respectively. Here, the Z_Y and Z_Z are impedances with very high values. While $\alpha(s)$ and $\beta(s)$ represent the unity current (from port X to port Z) and voltage (from port Y to port X) transfer functions, ideally $\alpha(s)$ and $\beta(s)$ are equal to 1. As shown in Figure 1, the value of these impedances, in particular Z_X (with a very small value), can be changed by varying the DC biasing current I_o of the conveyor, giving rise to the concept of second generation current controlled current conveyor (CCCII).

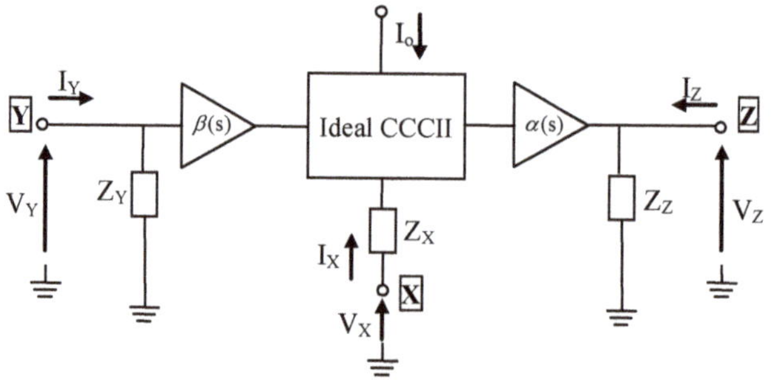

Figure 1. Equivalent circuit of CCCII.

The governing equation of the CCCII can be given in matrix form as:

$$\begin{bmatrix} I_Y \\ V_X \\ I_Z \end{bmatrix} = \begin{bmatrix} \frac{1}{Z_Y(I_o)} & 0 & 0 \\ \beta(s) & Z_X(I_o) & 0 \\ 0 & \alpha(s) & \frac{1}{Z_Z(I_o)} \end{bmatrix} \cdot \begin{bmatrix} V_Y \\ I_X \\ V_Z \end{bmatrix} \quad (1)$$

The CCCII acts as a current follower (between port X and port Z), a voltage follower (between port Y and port X), and a transconductor (between port Z and port Y). The possibility of controlling Z_X (mainly consists of the parasitic resistance R_X) using the DC biasing current I_o, has made it to extend the applications of CCCIIs to the domain of controlled electronic functions.

2.2. The CCCII-Based LNA Principle of Operation

The CCCII is an increasingly popular building block for RF functions [20–22]. The CCCII can be used to design a gain-controllable LNA, the gain of which can be controlled by the DC biasing current I_o. For the realization of this LNA, three different CCCIIs, as the main building blocks, which are symbolized as CCCII 1 to CCCII 3, can provide voltage signal amplification and wideband output impedance matching when connected, as shown in Figure 2. The current conveyor CCCII 1 converts the input voltage signal into a current signal, the connection of the two current conveyors, CCCII 1 and CCCII 2, amplify this current signal, and the current conveyor CCCII 2 converts the amplified current signal into the output voltage signal. Besides, the current conveyor CCCII 3 is the Z-match block on the output of the LNA, which adjusts the output impedance of the LNA to an ideal value.

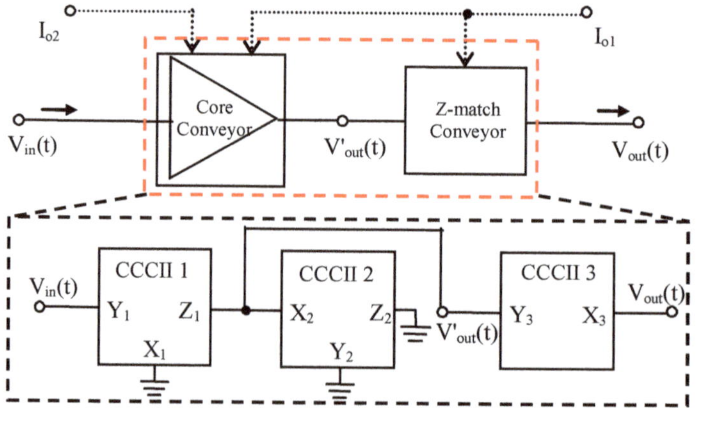

Figure 2. The operational principle of CCCII-based LNA.

The input voltage signal $V_{in}(t)$ is provided at port Y_1 of the current conveyor CCCII 1. Port Z_1 of current conveyor CCCII 1 and port X_2 of the current conveyor CCCII 2 are connected to each other, and the output voltage signal $V'_{out}(t)$ is tapped at this common connection. Assuming Z_{X1} and Z_{X2} to be purely resistive in nature (as L_{X1} and L_{X2} are the negligible parasitic inductances), with R_{X1} and R_{X2}, $V_{in}(t)$ is first converted by R_{X1} into $I_{X1}(t)$, which is given by

$$I_{X1}(t) = -\frac{V_{in}(t)}{R_{X1}} \tag{2}$$

$I_{X1}(t)$ is copied to port Z_1 as $I_{Z1}(t)$ (as to the current conveyor CCCII 1 between its port X_1 and port Z_1 is a current follower). Because of the connection of the two current conveyors CCCII 1 and CCCII 2, this current signal I_{Z1} comes into port X_2 of the current conveyor CCCII 2. Therefore, the current signal $I_{X1} = I_{Z1} = I_{X2}$ can be converted to a voltage signal owing to the parasitic resistance R_{X2} of CCCII 2. The resulting output voltage signal $V'_{out}(t)$ is given by

$$V'_{out}(t) = -R_{X2} \cdot I_{X1}(t) = \frac{R_{X2}}{R_{X1}} \cdot V_{in}(t) \tag{3}$$

In a typical CCCII operation, the parasitic resistance R_X at port X is inversely proportional to the DC biasing current I_o of the current conveyor, which is the fundamental property of the CCCII [20]. Then, the voltage gain (G) of the designed LNA is given by

$$G = \frac{V'_{out}(t)}{V_{in}(t)} = \frac{R_{X2}}{R_{X1}} = \frac{I_{o1}}{I_{o2}} \tag{4}$$

where R_{x1} and R_{x2} represent the parasitic resistances at port X of the corresponding current conveyor and I_{o1} and I_{o2} are their respective DC biasing currents. Under a certain DC biasing current I_{o2} range condition, the inverse relationship of voltage gain G to DC biasing current I_{o2} enables higher voltage gain at lower DC biasing current and, thus, at lower power consumption.

2.3. The CCCII-Based LNA Circuit Design

One of the base demands in the wideband RF LNA design is to have 50 Ohm input impedance matching [5]. According to the schematic as shown in Figure 2, the input signal is fed at port Y_1 of CCCII 1, at which intrinsic impedance Z_{Y1} (R_{Y1} in series with C_Y) at port Y_1 is of some tens or hundreds of kOhm. However, port X_1 of CCCII 1 has impedance R_{X1} lower than some hundreds of Ohm. On the other hand, the impedance R_{X1} can be reduced to 50 Ohm for higher values of the DC biasing current I_{o1}. Hence, the input signal is changed from port Y_1 to port X_1. The following optimization of the circuit architecture leads to the schematic of the proposed CCCII-based LNA as presented in Figure 3.

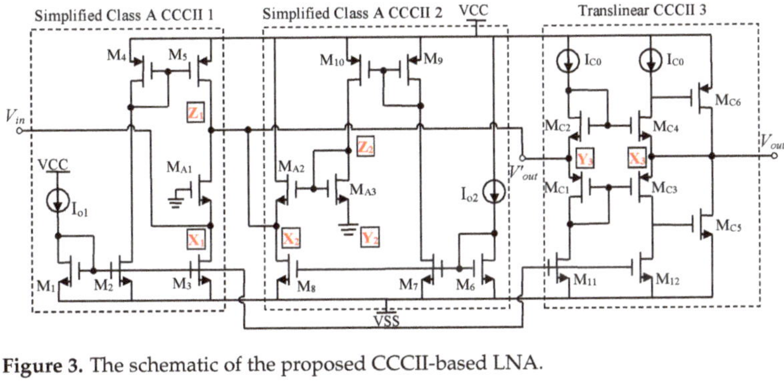

Figure 3. The schematic of the proposed CCCII-based LNA.

As can be seen from Figure 3, the proposed LNA utilizes two simplified class A CCCIIs [23] as building blocks to implement the amplifier stage. The current conveyor

CCCII 1 consists of an input common gate gain stage M_{A1} and the other current mirror transistors M_1~M_5, and the current conveyor CCCII 2 consists of a source follower stage M_{A2}–M_{A3} and the other current mirror transistors M_6~M_{10}. The output of the input transistor M_{A1} is connected to the input of the source follower M_{A2}. Note that the core of the proposed LNA comprises only three NMOS transistors (M_{A1}~M_{A3}) in its signal path, thereby allowing for a higher bandwidth to be obtained and improving the noise performance. The current sources I_{o1} and I_{o2} are used to bias the main NMOS transistors M_{A1} and M_{A2}–M_{A3}, as explained later, to adjust the voltage gain and to match the input impedance without the inductor components.

Based on a first order analysis in the wideband operation mode of Figure 3, a voltage gain can be given by

$$G(s) = \frac{V'_{out}(s)}{V_{in}(s)} = \frac{g_{m1}}{g_{m2} + C_T s} \tag{5}$$

where C_T is the total parasitic capacitance at the output node of port Z_1 with the input node of port X_2 and port Y_3, while g_{m1} and g_{m2} are the transconductances of M_{A1} and M_{A2}, respectively. This transfer function obtains a single dominant pole frequency, which approximately determines the amplifier bandwidth by

$$f_{-3dB} = \frac{g_{m2}}{2\pi C_T} \tag{6}$$

And the DC voltage gain approximately by

$$G = \frac{g_{m1}}{g_{m2}} \tag{7}$$

At the low frequency band, the input impedance can be approximately described by

$$Z_{in} = \frac{1}{g_{m1}} \tag{8}$$

Equations (6) and (7) show that a compromise on g_{m1} and g_{m2}, hence on the transistor sizes, as well as the DC biasing currents I_{o1} and I_{o2} of M_{A1} and M_{A2}, respectively, is necessary to optimize voltage gain and bandwidth. As can be seen, I_{o1} through g_{m1} affects the gain but has a little effect on the bandwidth, whereas I_{o2} controls both gain and bandwidth through g_{m2}. Therefore, varying I_{o2}, g_{m2} can be varied and the gain bandwidth product can be suitably adjusted. Increasing I_{o1} boosts the gain up again, but should be optimized to keep the power consumption sufficiently low. Under a certain DC biasing current I_{o2} range condition, the inverse relationship of voltage gain G to DC biasing current I_{o2} enables higher gain at lower DC biasing current and, therefore, at lower power consumption. Meanwhile, without the need of an LC matching network, Equation (8) indicates Z_{in} can be matched to the source impedance (50 Ohm) through I_{o1}. From Figure 3, the proposed wideband LNA is an inductorless circuit, its chip area is smaller than the most other LNAs.

In Figure 3, the number of components in the signal chain is low: only three NMOS transistors M_{A1}~M_{A3}. The main noise sources of the proposed LNA are contributed by the input NMOS transistor M_{A1}. When neglecting the parasitic capacitances and the noise of I_{o1} and I_{o2}, the noise figure of the LNA is approximately given by

$$NF = 1 + \gamma(1 + \frac{g_{m2}}{g_{m1}} + \frac{g_{m5}}{g_{m1}}) \tag{9}$$

The excess noise factor γ is a constant that depends on the transistor size, and g_{m5} represents the transconductance of M_5. As shown from Equation (9), the noise figure decreases when g_{m1} is increased. It can be achieved by increasing the transistor size of M_{A1} or the DC biasing current I_{o1}.

2.4. The Output Impedance Matching Design

In Figure 2, the Z-match block on the output is the impedance conversion circuit, which uses a translinear CCCII in the voltage follower mode to adjust the output impedance of the LNA to an ideal value. This design makes the output impedance of the LNA fully independent of its gain. In Figure 3, the core of translinear CCCII is composed of NMOS transistors (M_{C2} and M_{C4}) and PMOS transistors (M_{C1} and M_{C3}) as a translinear loop. At the same time, the transistors M_{C6} and M_{C4}, respectively, M_{C5} and M_{C3}, implement the other feedback loops to adjust the output impedance at port X_3 of the conveyor CCCII 3. The DC biasing current I_{o1} is copied to the required branches using current mirror transistors M_1 and M_{11}–M_{12}. The signal to be adapted, V'_{out}, is fed at port Y_3 of the current conveyor CCCII 3 ($V'_{out} = V_{Y3}$), which can be considered an open circuit since its intrinsic resistance, R_{Y3}, is very high. The output is tapped at port X_3 of the current conveyor CCCII 3 ($V_{out} = V_{X3}$). The voltage transfer function $\beta(s)$ (ideally equal to 1) of the voltage follower mode is

$$\beta(s) = \frac{V_{X3}(s)}{V_{Y3}(s)} = \frac{V_{out}(s)}{V'_{out}(s)} \approx 1 \tag{10}$$

The output impedance, at port X_3 of the conveyor CCCII 3, consists of resistance R_{X3} in series with a negligible parasitic inductance L_{X3}. By changing the DC biasing current I_{o1} of the conveyor, different resistance R_{X3} values can be obtained.

As can be seen in Figure 3, the CCCII-based output impedance matching circuit in this section has several advantages. Firstly, according to the Friis equation [24], the proposed impedance matching circuit is located at the output of the LNA, so its insertion has a little effect on the overall noise figure. Secondly, the gain-controllable characteristic of the LNA is not impacted by the insertion of the matching circuit, therefore there is no attenuation of the amplifying signal. Thirdly, the additional transistors in the signal path led to an increased immunity to reverse signal propagation. As a result, the insertion of the matching circuit led to the output impedance being perfectly matched to the ideal value (50 Ohm).

Compared to the traditional passive component solutions, the output impedance matching circuit of the proposed CCCII-based LNA is vastly superior by being: the small chip area without the inductor components, the flexible matching circuit to adjust any impedance to the ideal value, and a novel example of wideband output matching for the RF LNA applications.

3. Simulation Results and Discussion

The proposed wideband LNA has been realized using a GlobalFoundries' 0.18 μm single-poly six-metal CMOS technology and simulated using Cadence SpectreRF simulator. The GlobalFoundries' Design Rule (YI-093-DR001_Rev1V_1.8V-3.3V) and GlobalFoundries' Spice Model spec (yi093dr001_1v_00_20090731a) are used. During the design process, a very important step is the realization of the layout. To maximize RF performance, the layout was made by following classical design rules commonly used for analogue RF circuit synthesis (e.g., transistor matching, symmetric design, etc.). In order to validate the design approach, optimization of biasing currents and width of transistors with extensive DC analysis are simulated. The layout of the proposed wideband LNA is presented in Figure 4, which takes an active chip area, including testing pads of 0.42 mm × 0.23 mm. The absence of inductor components results in a very small chip area (0.096 mm^2).

Figure 5 presents the simulated input return loss (S11) with I_{o1} = 1.25 mA, I_{o2} = 50 μA, and V_{DC} = ±1.5 V. At a frequency of 1.0 GHz, S11 has a value of −21.4 dB and it remains below −16.6 dB for frequencies from 0.5 GHz to 2.5 GHz. As a critical parameter of the LNA, the S11 provides a good wideband input matching performance without the LC matching network. Figure 5 also presents the simulated output return loss (S22), which uses the translinear CCCII 3 in voltage follower mode as the output impedance matching circuit. The simulated S22 is from −10.6 dB to −18.6 dB over the entire frequency band of 0.5~2.5 GHz, indicating a reasonably acceptable output matching condition.

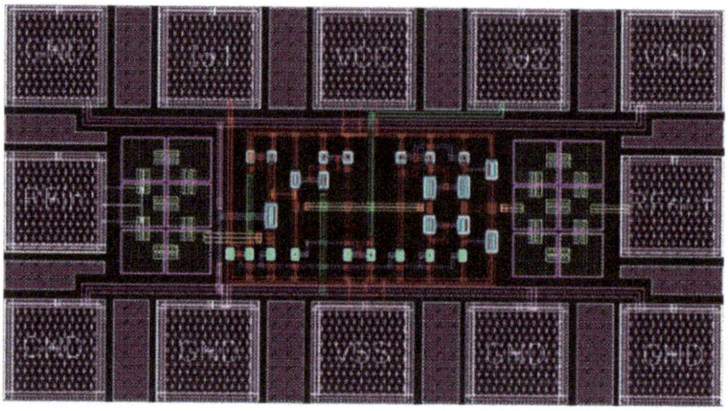

Figure 4. The layout diagram of the proposed wideband LNA.

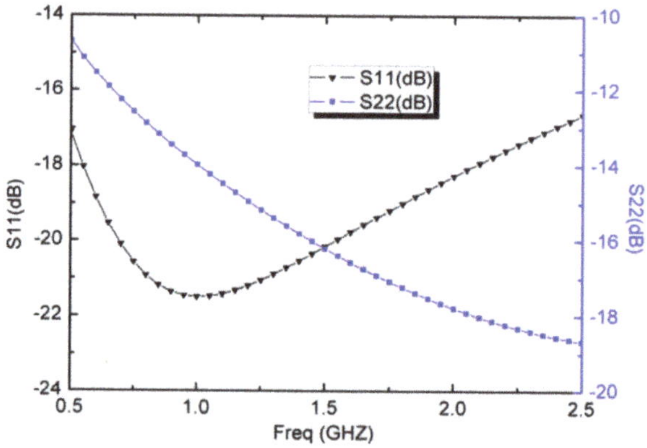

Figure 5. Simulated S11 and S22 of the wideband LNA.

In the proposed LNA, gain control over a wide range is achieved by varying the DC biasing current, I_{o2}, of the CCCII 2. Figure 6 shows that the voltage gain of the proposed LNA can be controllable between 18 dB and 1 dB when I_{o2} varies between 30 μA and 400 μA, at the same time, the LNA consumes between 3.8 mA and 5 mA for the entire range of I_{o2}. The inversely proportional relationship between the voltage gain and the DC biasing current indicates that better gain is achieved at the lower power consumption.

As described by Equations (6) and (7), the voltage gain increased and bandwidth decreased when I_{o2} decreased. In order to control the LNA performance over the 0.5~2.5 GHz bandwidth of interest, the currents I_{o1} and I_{o2} are fixed at 1.25 mA and 50 μA for this design. Under these conditions, the typical voltage gain (S21) and reverse isolation (S12) performances are present in Figure 7. The simulated S21 shows a typical voltage gain of 12.6 dB with a gain ripple of ±1.5 dB from 0.5 GHz to 2.5 GHz. According to Figure 7, the simulated S12 is from −65.2 dB to −39.5 dB in the required band, which indicates that the designed LNA can achieve a good stability characteristic. The better S12 can decrease the subsequent local oscillation (LO) leakage, arising from the substrate coupling and the capacitive paths.

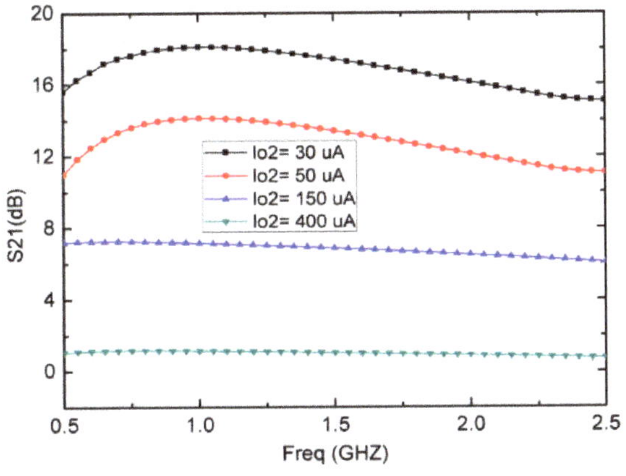

Figure 6. Simulated gain controllable range for various I_{o2} with I_{o1} = 1.25 mA and V_{DC} = ±1.5 V.

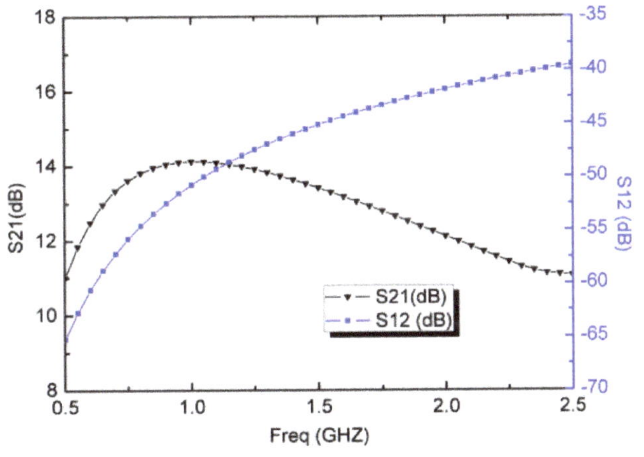

Figure 7. Simulated S21 and S12 of the wideband LNA.

As shown in Figure 8, the proposed wideband LNA is simulated for different temperatures and corner processes. The wideband LNA designed here has an excellent thermal stability: in the temperature range of −25 °C to 75 °C, the gain S21 drops about by ±1 dB. As the temperature increases, the gain S21 of the wideband LNA gradually decreases.

Owing to noise concerns, Figure 9 shows the simulated NF of the wideband LNA with I_{o1} = 1.25 mA and I_{o2} = 50 μA. It has a relatively flat and low NF performance across the entire 2 GHz bandwidth. From Figure 9, an excellent NF of 4.05~4.35 dB in the frequency range of 0.5~2.5 GHz is achieved. The maximum NF is 4.35 dB at 0.5 GHz and the minimum NF is 4.05 dB at 1.5 GHz, with an average NF of 4.2 dB.

To observe the nonlinear behavior, the two-tone signals at 1.50 GHz and 1.51 GHz with equal power levels are applied to the wideband LNA. The two-tone signals with 10 MHz spacing are used to simulate the fundamental and third-order inter-modulation (IM3) output power, versus input power, characteristics. As shown in Figure 10, an input third-order intercept point (IIP3) of −2.5 dBm and an input 1 dB compression point of −12.5 dBm are achieved. These results show that the designed LNA can achieve good linearity even with the low NF.

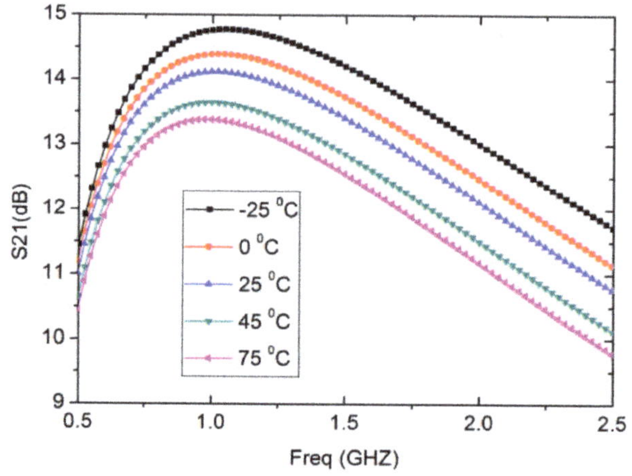

Figure 8. Simulated S21 of the wideband LNA with temperature variations.

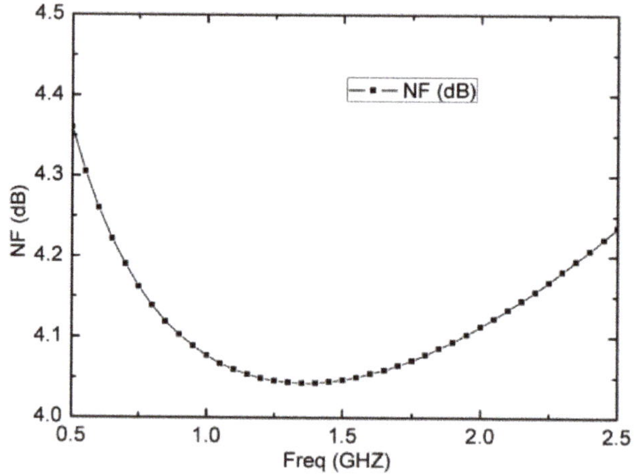

Figure 9. Simulated noise figure of the wideband LNA.

For I_{o2} = 50 μA, and at a dual supply voltage of ±1.5 V, the CCCII-based LNA consumes a current of 4 mA which results in a total power consumption of 12 mW. Finally, Table 1 is a summary of the proposed CMOS 0.5~2.5 GHz wideband LNA and reports recent CMOS state-of-the-art wideband LNA designs. Compared to other wideband LNA designs, the post-layout simulation results show that the proposed LNA can achieve a good wideband input/output impedance matching by using the CCCIIs, and further benefits include its inductorless design with smaller chip area and the capability of gain-controllability by varying a DC biasing current (I_{o2}).

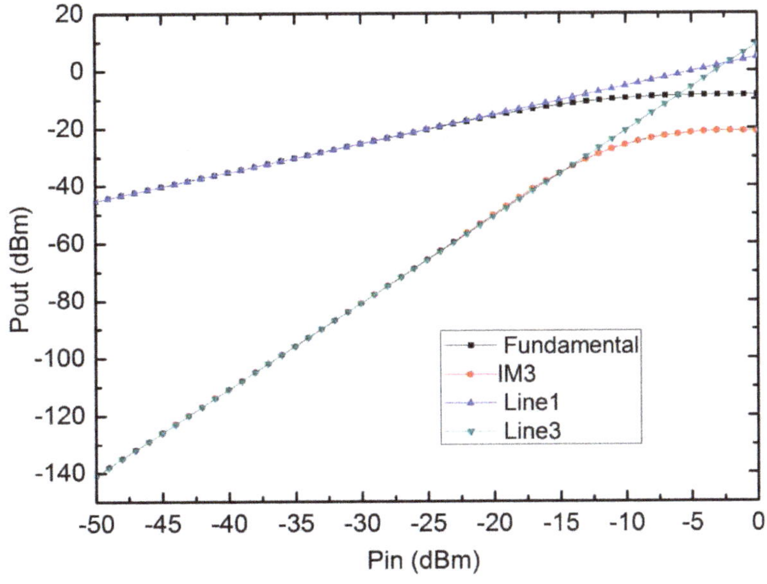

Figure 10. Simulated IIP3 and P1dB of the wideband LNA.

Table 1. Performance summary of the proposed LNA and comparison with other designs.

Specification	[6] S	[7] S	[8] M	[13] S	[14] M	This Work S
Technology	65 nm CMOS	0.18 μm CMOS	0.13 μm CMOS	0.18 μm CMOS	0.18 μm CMOS	0.18 μm CMOS
Frequency (GHz)	0.5~5	2.3~4.8	3.1~4.8	2~3.3	3.7~11.9	0.5~2.5
Input return loss S11 (dB)	<−10	<−10	<−8.4	<−10	<−8.5	<−16.6
Output return loss S22 (dB)	<−10	–	<−14	–	<−10	<−10.6
Voltage gain maximum (dB)	13	24	13	16.8	8.1	14.1
Reverse isolation S12 (dB)	–	–	<−40	–	<−30.3	<−39.5
NF (dB)	4~4.5	2.8–3.7	3.5~4.5	5.55~6	3.02~3.86	4.05~4.35
IIP3 (dBm)	−10	−3.2	−6.1	0.5	−1.3	−2.5
Power dissipation (mW)	5.0 #	13.1	3.4	2.55 #	7.3	12.0
Die size (mm^2)	0.173	0.34	0.40	0.283	0.34	0.096

Without the output buffer dissipation. S The post-layout simulation result. M The measurement result.

4. Conclusions

In this paper, a new wideband LNA which utilizes current conveyors as building blocks has been presented. Thanks to the use of three different CCCIIs for conveying the signal, the proposed wideband LNA has the following notable advantages over recent works: the total absence of inductor components, and thus, the LNA has an active chip area of only 0.096 mm^2; wideband input/output impedance matching and stable frequency responses in the required 0.5~2.5 GHz band; easily gain-controllable over a wide range (from 1 dB to 18 dB). The above excellent performances have shown that the proposed CCCII-based wideband LNA is suited to the software-defined and multi-standard wireless communication systems.

Author Contributions: Conceptualization, Q.W.; methodology, Q.W.; software, J.L. and S.C.; validation, Q.W., J.L. and S.C.; formal analysis, J.L.; investigation, S.C.; resources, Q.W.; writing—original draft preparation, Q.W.; writing—review and editing, Q.W.; visualization, J.L. and S.C.; supervision, Q.W.; project administration, Q.W.; funding acquisition, Q.W. All authors have read and agreed to the published version of the manuscript.

Funding: This research was supported by the National Natural Science Foundation of China (Grant No. 61901169) and the Natural Science Foundation of Hunan Province, China (Grant No. 2019JJ40190).

Data Availability Statement: The data that support the findings of this study are available within the article.

Acknowledgments: The authors would like to thank the anonymous reviewers and editor for their valuable comments which helped in improving this manuscript.

Conflicts of Interest: The authors declare no conflict of interest.

References

1. Hussain, I.; Wu, K. Concurrent Dual-Band Heterodyne Interferometric Receiver for Multistandard and Multifunction Wireless Systems. *IEEE Trans. Microw. Theory Tech.* **2021**, *69*, 4995–5007. [CrossRef]
2. Kanchetla, V.K.; Kharalkar, A.; Jain, S.; Khyalia, S.K.; Joy, J.; Jose, S.; Zele, R. A Compact, Reconfigurable CMOS RF Receiver for NavIC/GPS/Galileo/BeiDou. *IEEE Trans. Microw. Theory Tech.* **2022**, *70*, 3361–3373. [CrossRef]
3. Wan, Q.Z.; Yan, Z.D.; Li, F.; Liu, J.; Chen, S.M. Multistable dynamics in a Hopfield neural network under electromagnetic radiation and dual bias currents. *Nonlinear Dyn.* **2022**, *109*, 2085–2101. [CrossRef]
4. Wan, Q.Z.; Yan, Z.D.; Li, F.; Chen, S.M.; Liu, J. Complex dynamics in a Hopfield neural network under electromagnetic induction and electromagnetic radiation. *Chaos* **2022**, *32*, 073107. [CrossRef] [PubMed]
5. Parvizi, M.; Allidina, K.; El-Gamal, M.N. An ultra-low-power wideband inductorless CMOS LNA with tunable active shunt-feedback. *IEEE Trans. Microw. Theory Tech.* **2016**, *64*, 1843–1853. [CrossRef]
6. Mauludin, M.F.; Lee, D.-H.; Kim, J. A Wideband Low-Power Balun-LNA with Feedback and Current Reuse Technique. *Electronics* **2022**, *11*, 1372. [CrossRef]
7. Akbar, F.; Atarodi, M.; Saeedi, S. Design method for a reconfigurable CMOS LNA with input tuning and active balun. *AEU—Int. J. Electron. Commun.* **2015**, *69*, 424–431. [CrossRef]
8. Khurram, M.; Hasan, S.M.R. A 3–5 GHz current-reuse gm-boosted CG LNA for ultrawideband in 130 nm CMOS. *IEEE Trans. Very Large Scale Integr. (VLSI) Syst.* **2012**, *20*, 400–409. [CrossRef]
9. Cen, M.C.; Song, S.X. A high gain, low-power low-noise amplifier for ultra-wideband wireless systems. *Circuits Syst. Signal Process.* **2014**, *33*, 3251–3262. [CrossRef]
10. Ghadiri, A.; Moez, K. Gain-enhanced distributed amplifier using negative capacitance. *IEEE Trans. Circuits Syst. I Regul. Pap.* **2010**, *57*, 2834–2843. [CrossRef]
11. Huang, J.-C.; Chen, P.; Kao, P.-C.; Wang, H. A novel distributed amplifier with high gain, low noise and high output power in 0.18-μm CMOS technology. *IEEE Trans. Microw. Theory Tech.* **2013**, *61*, 1533–1542.
12. Saberkari, A.; Shirmohammadli, V.; Yagoub, M.C.E. A 3–6 GHz Current Reused Noise Canceling Low Noise Amplifier for WLAN and WPAN Applications. *Wirel. Pers. Commun.* **2016**, *86*, 1359–1376. [CrossRef]
13. Atakan, A.; Deniz, T.; Okan, Z.B.; Gunhan, D. A 2.55-mW on-chip passive balun-LNA in 180-nm CMOS. *Analog Integr. Circuits Signal Process.* **2022**, *111*, 223–234.
14. Chang, J.F.; Lin, Y.S. Complementary Current-Reused 3.7–11.9 GHz LNA Using Body-Floating and Self-Bias Technique for Sub-6 GHz 5G Communications. *Circuits Syst. Signal Process.* **2022**, *41*, 5968–5989. [CrossRef]
15. Jin, T.H.; Han, H.G.; Kim, T.W. A 0.7-dB NF, +8.2-dBm IIP3 CMOS low noise amplifier using frequency selective feedback. *Int. J. Circuit Theory Appl.* **2016**, *44*, 21–37. [CrossRef]
16. Koushick, M.; Sandhya, P.; Rabindranath, N. Electronically Tunable Dual-Input Integrator: Synthetic Immittance Function Realizability. *IETE J. Res.* **2017**, *63*, 713–718.
17. Rana, C.; Prasad, D.; Afzal, N. High-performance programmable grounded resistor and its applications. *Automatika* **2018**, *59*, 71–77. [CrossRef]
18. Godara, B.; Fabre, A. A highly compact active wideband balun with impedance transformation in SiGe BiCMOS. *IEEE Trans. Microw. Theory Tech.* **2008**, *56*, 22–30. [CrossRef]
19. Arslan, E.; Morgül, A. Self-Biasing Current Conveyor for High Frequency Applications. *J. Circuits Syst. Comput.* **2012**, *21*, 1250039. [CrossRef]
20. Abbas, Z.; Olivieri, M.; Yakupov, M.; Ripp, A. Design centering/yield optimization of power aware band pass filter based on CMOS current controlled current conveyor (CCCII+). *Microelectron. J.* **2013**, *44*, 321–331. [CrossRef]
21. Seguin, F.; Godara, B.; Alicalapa, F.; Fabre, A. A gain-controllable wide-band low-noise amplifier in low-cost 0.8 um Si BiCMOS Technology. *IEEE Trans. Microw. Theory Tech.* **2004**, *52*, 154–160. [CrossRef]

22. Jin, J.; Zhou, K.; Zhao, L. Designing RF Ring Oscillator Using Current-Mode Technology. *IEEE Access* **2017**, *5*, 5306–5312. [CrossRef]
23. Seguin, F.; Fabre, A. 2 GHz controlled current conveyor in standard 0.8 μm BiCMOS technology. *Electron. Lett.* **2001**, *37*, 329–330. [CrossRef]
24. Abidi, A.A.; Leete, J.C. De-embedding the noise figure of differential amplifiers. *IEEE J. Solid-State Circuits* **1999**, *34*, 882–885. [CrossRef]

micromachines

MDPI

Article

Experimental Investigation of Relationship between Humidity Conditions and Degradation of Key Specifications of 0.1–1.2 GHz PA in 0.18 μm CMOS

Shaohua Zhou [1,2,†], Cheng Yang [1,2,†] and Jian Wang [1,2,*]

[1] School of Microelectronics, Tianjin University, Tianjin 300072, China; zhoushaohua@tju.edu.cn (S.Z.); ych2041@tju.edu.cn (C.Y.)
[2] Qingdao Institute for Ocean Technology, Tianjin University, Qingdao 266200, China
* Correspondence: wangjian16@tju.edu.cn
† These authors contributed equally to this work.

Abstract: The specification of power amplifiers (PA) is closely related to humidity variation, and few reports on the humidity properties of PA are available in the literature. Therefore, an experimental study of PA specifications was conducted under different humidity conditions to elucidate the relationship between the degradation of PA specifications and humidity conditions. This paper studies and provides results of the degradation of a PA subjected to different humidity levels. The experimental results show that the S_{21} and output power decrease with the increase in humidity. The main cause of this degradation is the decrease in oxide capacitance and increase in threshold voltage with increasing humidity, resulting in a reduction of transconductance and an increase in on-resistance. The results of this study can guide designers in designing compensation circuits to achieve some degree of compensation for the degradation of PA specifications.

Keywords: class-A; CMOS; humidity; performance degradation; power amplifier (PA)

Citation: Zhou, S.; Yang, C.; Wang, J. Experimental Investigation of Relationship between Humidity Conditions and Degradation of Key Specifications of 0.1–1.2 GHz PA in 0.18 μm CMOS. *Micromachines* **2022**, *13*, 1162. https://doi.org/10.3390/mi13081162

Academic Editor: Zhiming Chen

Received: 12 June 2022
Accepted: 21 July 2022
Published: 22 July 2022

1. Introduction

With the rapid expansion in the wireless industry [1,2], the power amplifiers (PAs), which play critical roles in developing wireless devices and establishing communication [3–5], are in high demand. For long-distance transmission, the antennas are driven mainly by the PAs [6–8]. The PAs employed in such applications are exposed to environmental conditions [9,10] and sometimes extreme working conditions such as rapid changes in temperature and humidity [11–14]. According to the literature, humidity caused 10% of equipment failures at US coastal bases [15]. The performance of the PA can be affected by the humidity conditions that may result in system failures [16,17]. A failure is an event that ends the life of a particular product [18,19]. However, we never use PA in existing communication systems until its end of life. Before reaching the end of a PA's lifetime, that PA's performance has already begun to degrade and cannot meet the required specifications of the designed systems [20,21].

From a systems point of view, the system cannot tolerate a drop in PA's specification, such as gain, bandwidth, etc. [19]. Once the performance of the PA is degraded, even if the magnitude of performance degradation is very small, it can lead to catastrophic damage to the overall system [22–24]. This means that when the performance of the PA degrades below a predetermined threshold, even if the degradation is minor, the PA is incapable of meeting the design specification [20,21]. Therefore, figuring out the PA's degradation concerning humidity variations becomes crucial to ensure the proper system operation. Unfortunately, the characteristics of gallium nitride (GaN)-based PAs in a humid environment have limited studies in the literature [20], while studies of Complementary Metal Oxide Semiconductor (CMOS)-based PAs exposed to varied humidity conditions are entirely lacking.

This paper uses a CMOS Class-A PA to systematically investigate the PA's specifications degradation under different humidity levels from an experimental point of view. The PA's specifications degrade as humidity rises. Furthermore, the PA's degradation mechanism under high humidity conditions is elucidated. This paper aims to help PA designers consider the humidity effect of PA in advance in the design process.

2. Designed PA and Experimental Setup

The 0.1–1.2 GHz CMOS broadband PA is a technological achievement result of the national science and technology project. The project requires the design of a broadband PA that can be used in industry-specific networks, and the study of the temperature and humidity characteristics of the PA is required to lay the foundation for future product finalization. Therefore, we conducted the humidity characteristics study with this PA in this context. The PA was processed in Global Foundry 1P6M 180 nm CMOS (The Foundry Company, Sunnyvale, CA, USA) with a chip area of 0.414 mm^2 [25], as shown in Figure 1. As you can see from Figure 1, there are ten pads in the chip, five of which are ground pads, three are power pads, and two are input and output pads. The supply voltage of V_{dd1}, V_{dd2}, and V_{dd3} is 3.3 V. Special attention should be paid to the fact that the chip ground (GND) pad must be connected to the real earth; otherwise, it will cause the test to fail, or the test result will be abnormal.

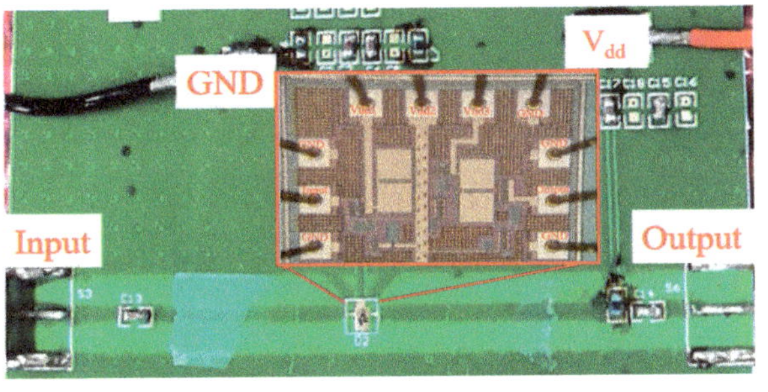

Figure 1. The chip microphotograph of the CMOS PA.

As shown in the circuit diagram of Figure 2, a PA with a three-stage single-ended structure was used for the experiments, and the first and second stages of the PA were cascaded to increase the gain of the PA. In the third stage, a common source amplifier + resistive negative feedback structure is used to achieve high power output. The PA uses a 3.3 V powered thick oxide N-channel metal-oxide semiconductor (NMOS) transistor in a 0.18 μm CMOS process to increase output power, with a minimum gate length of 0.35 μm.

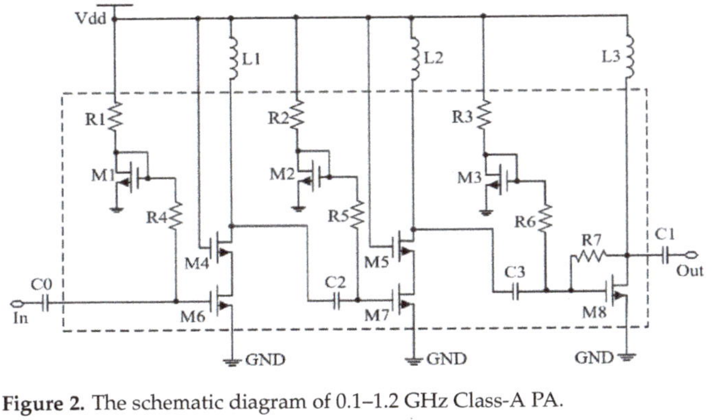

Figure 2. The schematic diagram of 0.1–1.2 GHz Class-A PA.

This experiment must be performed in a Solar Climatic Test Cabinet (SC^3 1000 MHG) (Vötsch Industrietechnik Gmbh, Balingen, Germany) (Figure 3). Unfortunately, the existing on-chip test platform does not support humidity experiments. Therefore, the chip to be tested can only be packaged on a printed circuit board (PCB) board (as shown in Figure 1) and then placed in the environmental test chamber for measurement. The PCB board used in this experiment is Rogers 4350 (Rogers, Killingly, CT, USA). According to the datasheet of the Murata inductor and capacitor (Murata, Nagaokakyō, Japan) and Rogers 4350, the performance of the inductor, capacitor, and PCB can be considered constant. Therefore, the PA's specifications degradation in this experiment can be mainly caused by the changes in the transistors in the PA. This way, the transistors' measurement in a humidity-changing environment is achieved indirectly.

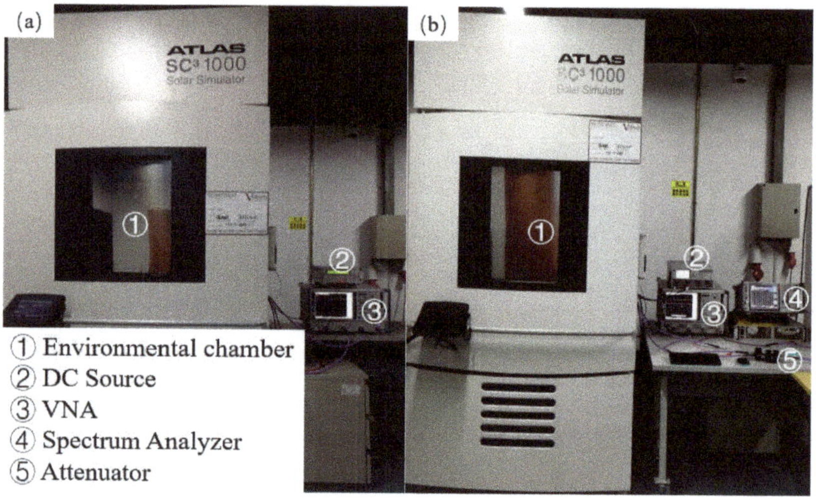

Figure 3. Measurement environment: (**a**) S-parameter; (**b**) output power.

To study the humidity characteristics of individual transistors, existing on-chip test benches can also fail to provide a humidity measurement environment. Therefore, individual transistors are packaged on a PCB and measured in an environmental test chamber. However, the transistor's input/output impedance is not the typical 50 Ω of the standard test system. Therefore, measuring a single transistor's performance under humidity variation is impossible using the standard test environment. Therefore, this paper does not discuss the characteristics of transistors in different humidity environments by directly measuring them.

The R&S HMP4040 (R&S, Munich, Germany) was used to provide DC bias, an attenuator (Rosenberger, Tittmoning, Germany) was used to reduce the power at the spectrum analyzer's input, and the R&S ZVA40 (R&S, Munich, Germany) and FSV30 (R&S, Munich, Germany) were used to record the experimental output results. Since the humidity range that the environmental test chamber can measure is 10% to 90% Relative Humidity (*RH*), this paper reports the variation of the S_{21} and output power in the humidity range of 10% to 90% *RH* only.

3. Results and Discussions

3.1. The S_{21}

The measured S_{21} is as in Figure 4. The measured S_{21} is not only the function of frequencies but also varies according to different humidity conditions. For example, at 500 MHz, the gain difference between 10% to 90% *RH* is 0.21 dB, but at 900 MHz, the difference is 0.32 dB. The differences in gain for the same humidity conditions will be different at different operating frequencies.

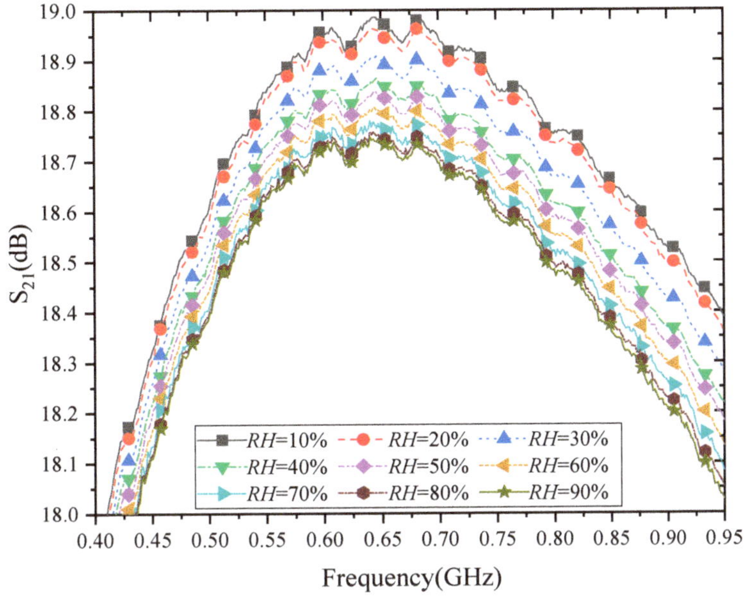

Figure 4. Measured small-signal gain with frequency variations.

Figure 5 gives the details of the humidity dependence of S_{21}. In this figure, we took 433 MHz and 900 MHz as examples based on two common communication methods: interphone and public mobile phone. If we take S_{21} at 20% *RH* as a typical value and 0.25 dB drop as an acceptable criterion, the maximum operating humidity is around 70% *RH*. This means that the gain of the power amplifier can meet the requirements below 70% *RH*. However, as shown in the introduction, if the S_{21} of the PA degrades is below 18.25 dB (e.g., the degradation is minor), the S_{21} of the PA will not be able to meet the demand at this time.

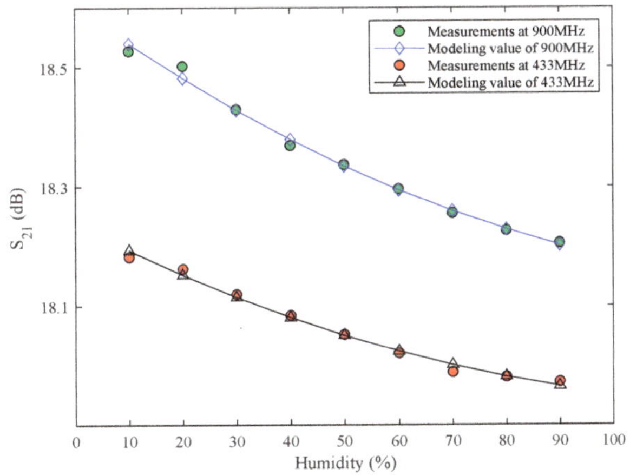

Figure 5. Measured S_{21} vs. humidity.

As shown in Figures 4 and 5, S_{21} of PA decreases with the increase of humidity. What is the cause of the degradation of S_{21} of PA with humidity? Next, this paper will discuss and analyze it in detail. For Class A, the PA operates at bias points $V_{gs\text{-}M8}$ = 1.7 V, $V_{ds\text{-}M8}$ = 3.3 V, and $I_{ds\text{-}M8}$ = 60 mA. Therefore, the PA works in the saturation region.

In the saturation region, we obtain [26–28]:

$$I_{ds} = \frac{W_g \mu_n C_{ox}}{2L_g}(V_{gs} - V_{th})^2 \tag{1}$$

where W_g is the gate width, μ_n is the carrier mobility, C_{ox} is the gate oxide capacitance per unit area, L_g is the gate length, V_{gs} is the gate voltage, V_{th} is the threshold V_{ds} is drain voltage.

The gate oxide capacitance per unit area is found to be [26–28]:

$$C_{ox} = \frac{\varepsilon_{ox}}{t_{ox}} \tag{2}$$

where ε_{ox} is the permittivity of the oxide, and t_{ox} is the thickness of the oxide.

As reported in the literature [29,30], the accumulation of water diffusion through the polysilicon sidewalls at the oxide/silicon interface near the edge of the source/drain junction will increase the thickness of the oxide. Therefore, the drain current and gate oxide capacitance per unit area will decrease with increasing oxide thickness according to Equations (1) and (2).

In addition to the possible reasons mentioned above, another reason for the decrease of the drain current is the increase of the threshold voltage with the rise in the humidity. This experimental work's finding aligns with the literature's theoretical confirmation [31].

The transconductance of a MOSFET is defined as the change in drain current concerning the corresponding change in gate voltage [26–28]

$$g_m = \frac{\partial I_{ds}}{\partial V_{gs}} \tag{3}$$

Therefore, according to Equation (3), the transconductance of the saturation region is [26–28]

$$g_{ms} = \frac{W_g \mu_n C_{ox}}{L_g}(V_{gs} - V_{th}) \tag{4}$$

It is already known from the previous discussion that the oxide capacitance (C_{ox}) decreases, and the threshold voltage increases as humidity rises. Therefore, according to Equation (4), it is known that the transconductance will decrease with the increase of humidity. And the transconductance is generally considered to be S_{21}. Therefore, it can be observed that S_{21} decreases with the rise in humidity.

In summary, there are two reasons for the degradation of S_{21}. The first reason is the increase in threshold voltage with increasing humidity. The second reason is that C_{ox} decreases with increasing humidity. It may be noted that these reasons were not observed directly by the experiment earlier.

In addition, we also fitted the relationship between S_{21} and humidity variation at 433 MHz and 900 MHz (as shown in Figure 5) based on the quadratic polynomial model, and the expressions are:

$$S_{21\text{–}433\text{MHz}} = 1.7765 \times 10^{-5} RH^2 - 4.6327 \times 10^{-3} RH + 18.2375 \tag{5}$$

$$S_{21\text{–}900\text{MHz}} = 2.2451 \times 10^{-5} RH^2 - 6.4763 \times 10^{-3} RH + 18.6023 \tag{6}$$

where RH is the humidity.

From Equations (5) and (6), the expressions of S_{21} with the temperature at 433 MHz and 900 MHz are formally the same, and both are quadratic functions of *RH*. Only the coefficients are different.

3.2. The PA's Output Power

We studied the characteristics of the output power of the power amplifier at 433 MHz, as this is a typical radio frequency for different humidity levels. The humidity characteristics of the output power as shown in Figure 6. It can be seen from it that output power decrease with increasing humidity. And when the humidity changes the same, the magnitude of the degradation of output power at different input power is not the same. As discussed in sub-Section 3.1, it is already known that the decrease in C_{ox} and the increase in threshold voltage cause the drain current to decrease with increasing humidity. Therefore, the following analysis focuses on the causes of the decrease in output power with increasing humidity.

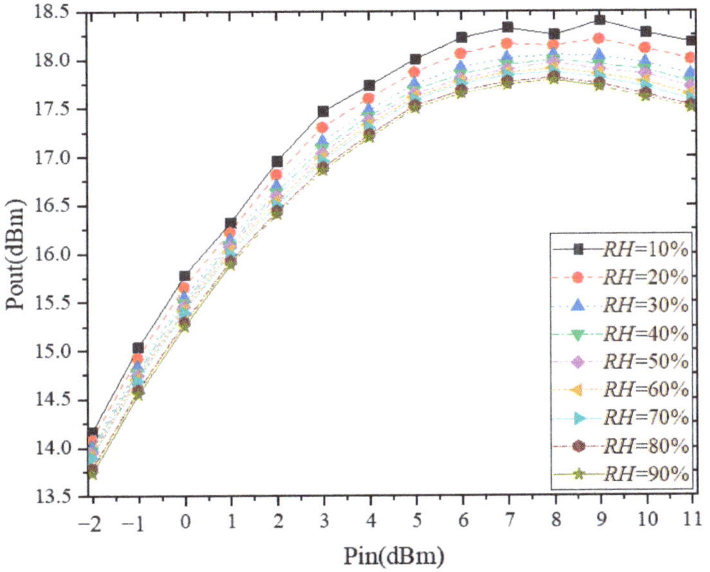

Figure 6. Measured output power vs. humidity at 433 MHz.

In the following, we will discuss and analyze specifically the causes of the decrease in output power with increasing humidity. The on-resistance is [32]:

$$R_{on} = R_0 + \frac{kT^{\alpha}}{|V_{gs} - V_{th}|^{\beta}} \tag{7}$$

Here, k is a constant, and R_0 represents the initial value of the resistance. The coefficients α and β are 1.5 and 0.2, respectively.

According to Equation (7), the on-resistance will increase with an increasing threshold voltage. As known from the discussion in 3.1, it has been demonstrated experimentally and in the literature that the threshold voltage increases with increasing humidity. This means that the on-resistance will increase with increasing humidity. The results of our experimental analysis are the same as in the literature [31]. Therefore, an increase in the on-resistance will lead to a decrease in the output power.

Figure 7 gives the relationship between saturation output power and humidity at 433 MHz, from which the saturation output power decreases with the increase of humidity.

At the same time, we fit the relationship of saturation output power with humidity based on the quadratic polynomial model, and the expression is:

$$P_{out}(sat) = 8.7944 \times 10^{-5} RH^2 - 1.5564 \times 10^{-2} RH + 18.5010 \quad (8)$$

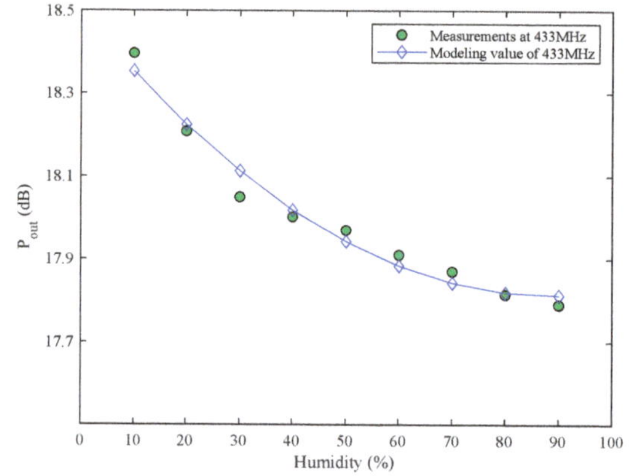

Figure 7. Saturation output power vs. humidity at 433 MHz.

From Equation (8), the saturation output power ($P_{out}(sat)$) with humidity is similar to the expression of S_{21} with humidity, and both are related to quadratic humidity. Only the specific coefficients are different.

Since we only measured the output power at 433 MHz during the previous measurements, the output power at 900 MHz is not provided.

3.3. The Power Added Efficiency

The relationship between the power added efficiency (PAE) and humidity at 433 MHz is shown in Figure 8, from which PAE decreases with the increase of humidity. According to the definition of PAE, PAE is proportional to the output power; i.e., PAE decreases as output power decreases. The reason for the change of output power with the humidity has been discussed in sub-Section 3.2, so the reason for the reduction of PAE with the increase of humidity is the same as the reason for the decrease of output power with the rise in humidity.

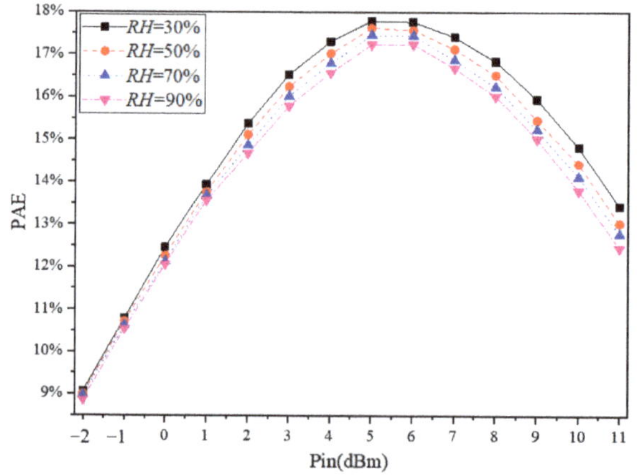

Figure 8. Measured PAE vs. humidity at 433 MHz.

It should be noted that considering that the curves of S_{21} and output power changing with humidity in Figures 4 and 6 are too dense, only PAE under four humidities are given in Figure 8 to observe the relationship between PAE and humidity change more clearly. Figure 9 shows the relationship between the maximum PAE and humidity at 433 MHz. At the same time, the relationship between the maximum PAE and humidity is fitted, and its expression is as follows

$$PAE_{max} = -9.0780 \times 10^{-3} RH + 18.0699 \quad (9)$$

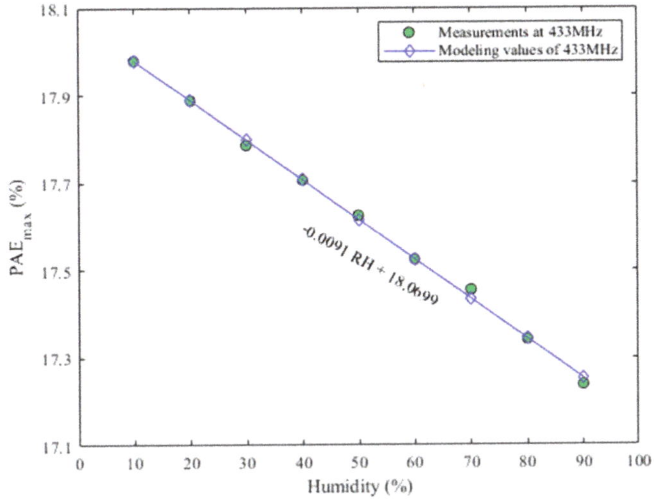

Figure 9. Measured maximum PAE vs. humidity at 433 MHz.

According to Equation (9), the maximum PAE is linear with the change of humidity, which is different from the relationship between S_{21} and saturated output power with humidity change.

4. Conclusions

This paper investigates PA specifications' characteristics at different humidity levels from an experimental point of view. This paper study and provide results of the degradation by the humidity of only one PA. It is found that the S_{21} and output power degrades with the increase of humidity. The main reason for this degradation is that the C_{ox} decreases, and the threshold voltage increases as humidity rises.

However, no common pattern for the degradation of the two specifications, S_{21}, and the output power, was found during the study. This indicates that the specification degradation experiments must be continued at this stage for many PAs to extract sufficient experimental data. This will guide the extraction of common patterns of specification degradation for the prediction of specification degradation and compensation of specification degradation.

Author Contributions: Conceptualization, S.Z., C.Y. and J.W.; methodology, S.Z., C.Y. and J.W.; software, S.Z. and C.Y.; validation, S.Z., C.Y. and J.W.; formal analysis, J.W.; investigation, S.Z. and C.Y.; resources, J.W.; data curation, S.Z. and C.Y.; writing—original draft preparation, S.Z.; writing—review and editing, C.Y. and J.W.; visualization, S.Z. and C.Y.; supervision, J.W.; project administration, J.W.; funding acquisition, J.W. All authors have read and agreed to the published version of the manuscript.

Funding: This research is supported by the State Key Laboratory of Complex Electromagnetic Environment Effects on Electronics and Information System (No. CEMEE2022G0201, CEMEE-002-20220224).

Institutional Review Board Statement: Not applicable.

Informed Consent Statement: Not applicable.

Data Availability Statement: Not applicable.

Conflicts of Interest: The authors declare no conflict of interest.

References

1. Jia, P.; You, F.; He, S. A 1.8–3.4-GHz Bandwidth-Improved Reconfigurable Mode Doherty Power Amplifier Utilizing Switches. *IEEE Microw. Wirel. Compon. Lett.* **2020**, *30*, 102–105. [CrossRef]
2. Del, D.; Gurutzeaga, I.; Beriain, A.; Solar, H.; Berenguer, R. A Compact, Wideband, and Temperature Robust 67–90-GHz SiGe Power Amplifier With 30% PAE. *IEEE Microw. Wirel. Compon. Lett.* **2019**, *29*, 351–353. [CrossRef]
3. Zurek, P.; Cappello, T.; Popovic, Z. Broadband Diplexed Power Amplifier. *IEEE Microw. Wirel. Compon. Lett.* **2020**, *30*, 1073–1076. [CrossRef]
4. Thayyil, M.V.; Li, S.; Joram, N.; Ellinger, F. A 4–32 GHz SiGe Multi-Octave Power Amplifier With 20 dBm Peak Power, 18.6 dB Peak Gain and 156% Power Fractional Bandwidth. *IEEE Microw. Wirel. Compon. Lett.* **2019**, *29*, 745–748. [CrossRef]
5. Chen, L.; Zhang, L.; Wang, Y. A 26.4-dB Gain 15.82-dBm 77-GHz CMOS Power Amplifier With 15.9% PAE Using Transformer-Based Quadrature Coupler Network. *IEEE Microw. Wirel. Compon. Lett.* **2020**, *30*, 78–81. [CrossRef]
6. Gashi, B.; Krause, S.; Rüdiger, Q.; Fager, C.; Ambacher, O. Investigations of active antenna doherty power amplifier modules under beam-steering mismatch. *IEEE Microw. Wirel. Compon. Lett.* **2018**, *28*, 930–932. [CrossRef]
7. Zhou, H.; Pal, A.; Mehta, A.; Nagasundaram, E.; Lees, J.; Mirshekar-syahkal, D. Power amplifier and antenna integration utilizing capacitively coupled feeding mechanism. In Proceedings of the 2017 IEEE International Symposium on Antennas and Propagation & USNC/URSI National Radio Science Meeting, San Diego, CA, USA, 9–14 July 2017; pp. 2355–2356.
8. Timoshenkov, V.; Chaplygin, Y.; Khlybov, A.; Rodionov, D.; Efimov, A. X-band power amplifier for active electronically scanned array antenna. In Proceedings of the 2018 IEEE Conference of Russian Young Researchers in Electrical and Electronic Engineering (EIConRus), Moscow and St. Petersburg, Russia, 29 January–1 February 2018; pp. 1337–1340.
9. Lin, X.; Seet, B.; Joseph, F. Wearable humidity sensing antenna for BAN applications over 5G networks. In Proceedings of the 2018 IEEE 19th Wireless and Microwave Technology Conference (WAMICON), Sand Key, FL, USA, 9–10 April 2018.
10. Wang, R.; Yin, W.Y.; Chen, Y.; Chen, J.; Zhou, L. Study on Harmonic Spur Characteristics of AlGaN/GaN HEMT PA at Different Temperatures. In Proceedings of the 2019 12th International Workshop on the Electromagnetic Compatibility of Integrated Circuits (EMC Compo), Hangzhou, China, 21–23 October 2019; pp. 290–292.
11. Pour, F.L.; Ha, D.S. A Temperature Compensated 5 GHz GaN on SiC Power Amplifier. In Proceedings of the 2020 IEEE 63rd International Midwest Symposium on Circuits and Systems (MWSCAS), Springfield, MA, USA, 9–12 August 2020; pp. 549–553.
12. Dhar, S.K.; Helaoui, M.; Ghannouchi, F.M. Temperature Dependent Robust Behavioral Modeling of Non-Linear Power Amplifier. In Proceedings of the 2018 Asia-Pacific Microwave Conference (APMC), Kyoto, Japan, 6–9 November 2018; pp. 378–380.
13. Lin, Y.C.; Zhong, J. A review of the influencing factors on anisotropic conductive adhesives joining technology in electrical applications. *J. Mater. Sci.* **2008**, *43*, 3072–3093. [CrossRef]
14. Zhou, S.H. Experimental investigation on the performance degradations of the GaN class-F power amplifier under humidity conditions. *Semicond. Sci. Technol.* **2021**, *36*, 035025. [CrossRef]
15. Talking about the 'Shelf Life' of Missiles. Available online: http://www.81.cn/jfjbmap/content/2018--03/09/content_201251.htm (accessed on 9 March 2018).
16. Lin, Q.; Fu, H.P.; Cheng, Q.F.; Zhu, Y.Y. Study of the index failure for power amplifier caused by temperature. *Analog Integr. Circuits Signal Process.* **2016**, *89*, 177–183. [CrossRef]
17. Kim, Y.T.; Kim, K.B.; Hyun, Y.E.; Kim, I.J.; Yang, S. Simulation study on the lifetime of electrochemical capacitors using the accelerated degradation test under temperature and voltage stresses. *Microelectron. Reliab.* **2015**, *55*, 2712–2720. [CrossRef]
18. Zhang, H. Reliability and lifetime prediction of LED drivers. In Proceedings of the 2017 14th China International Forum on Solid State Lighting: International Forum on Wide Bandgap Semiconductors China (SSLChina: IFWS), Beijing, China, 1–3 November 2017.
19. Ma, J.; Zhou, S. Experimentally Investigating the Performance Degradations of Power Amplifiers Against Temperatures. In Proceedings of the 2020 International Conference on Microwave and Millimeter Wave Technology (ICMMT), Shanghai, China, 20–23 September 2020.
20. Zhou, S.H. Experimentally investigating the degradations of small-signal gain for a GaN class-AB dual-band power amplifier under high temperature and humidity conditions. *AIP Adv.* **2020**, *10*, 125219. [CrossRef]
21. Zhou, S.H. Experimentally investigating the degradations of the GaN PA indexes under different temperature conditions. *Microw. Opt. Technol. Lett.* **2021**, *63*, 758–763. [CrossRef]
22. Kim, M.; Bae, S.; Choung, S.H. Analysis on reliability characteristics of power electronic devices for aerial vehicles. In Proceedings of the 2017 20th International Conference on Electrical Machines and Systems (ICEMS), Sydney, Australia, 11–17 August 2017.
23. Domginuez-Garcia, A.; Kassakian, J.; Schindall, J. Reliability evaluation of the power supply of an electrical power net for safety-relevant applications. *Reliab. Eng. Syst. Saf.* **2006**, *91*, 505–514. [CrossRef]
24. Islam, A.E. Current status of reliability in extended and beyond CMOS devices. *IEEE Trans. Device Mater. Reliab.* **2016**, *16*, 647–666. [CrossRef]

25. Wu, H.F.; Wang, L.G.; Zhou, P.; Ma, J.G. A 0.1–1.2 GHz CMOS ultra-broadband power amplifier. In Proceedings of the 2014 IEEE MTT-S International Microwave Symposium (IMS2014), Tampa, FL, USA, 1–6 June 2014.
26. Neamen, D.A. *Semiconductor Physics and Devices: BASIC Principles*; McGraw-Hill: New York, NY, USA, 2011; pp. 387–410.
27. Razavi, B. *Design of Analog CMOS Integrated Circuits*; McGraw Hill Higher Education: New York, NY, USA, 2003; pp. 17–21.
28. Lee, T.H. *The Design of CMOS Radio-Frequency Integrated Circuits*; Cambridge University Press: Cambridge, UK, 2012; pp. 172–189.
29. Lee, Y.; Hui, S.; Sabi, B. Evaluation of moisture-induced hot-carrier degradation of N- and P-MOSFETs using steam stress. In Proceedings of the IEEE International Electron Devices Meeting, Washington, DC, USA, 5–8 December 1993.
30. Jiang, C.; Hu, C.; Chen, C.H.; Tseng, P.N. Impact of inter-metal-oxide deposition condition on NMOS and PMOS transistor hot carrier effect. In Proceedings of the 30th Annual Proceedings Reliability Physics 1992, San Diego, CA, USA, 31 March–2 April 1992.
31. Kyriakidis, A. Performance Evaluation of SiC Power MOSFETs for Hybrid & Electric Vehicle DC-DC Converters. Master's Thesis, Escola Tècnica Superior d'Enginyeria Industrial de Barcelona, Barcelona, Spain, September 2018.
32. Magnone, P.; Petucco, A.; Thevenet, N.; Abedini, H. Simplified on-line monitoring system of MOSFET on-resistance based on a semi-empirical model. In Proceedings of the 2019 34th Annual IEEE Applied Power Electronics Conference and Exposition (APEC), Anaheim, CA, USA, 17–21 March 2019.

micromachines

MDPI

Review

A State-of-the-Art Review on CMOS Radio Frequency Power Amplifiers for Wireless Communication Systems

Sofiyah Sal Hamid [1], Selvakumar Mariappan [1,*], Jagadheswaran Rajendran [1], Arvind Singh Rawat [2], Nuha A. Rhaffor [1], Narendra Kumar [3], Arokia Nathan [4] and Binboga S. Yarman [5]

[1] Collaborative Microelectronics Design Excellence Centre (CEDEC), Universiti Sains Malaysia, Bayan Lepas 11900, Malaysia; sofiyah@usm.my (S.S.H.); jaga.rajendran@usm.my (J.R.); nuha@usm.my (N.A.R.)
[2] School of Computing, DIT University, Dehradun 248009, Uttarakhand, India; arvindsrawat@ieee.org
[3] Department of Electrical Engineering, Faculty of Engineering, University of Malaya, Kuala Lumpur 50603, Malaysia; narendra.k@um.edu.my
[4] Darwin College, Cambridge University, Cambridge CB3 9EU, UK; an299@cam.ac.uk
[5] Department of Electrical and Electronics Engineering, Istanbul University, 34320 Istanbul, Turkey; sbyarman@gmail.com
* Correspondence: selva_kumar@usm.my

Abstract: Wireless communication systems have undergone significant development in recent years, particularly with the transition from fourth generation (4G) to fifth generation (5G). As the number of wireless devices and mobile data usage increase, there is a growing need for enhancements and upgrades to the current wireless communication systems. CMOS transceivers are increasingly being explored to meet the requirements of the latest wireless communication protocols and applications while achieving the goal of system-on-chip (SoC). The radio frequency power amplifier (RFPA) in a CMOS transmitter plays a crucial role in amplifying RF signals and transmitting them from the antenna. This state-of-the-art review paper presents a concise discussion of the performance metrics that are important for designing a CMOS PA, followed by an overview of the trending research on CMOS PA techniques that focuses on efficiency, linearity, and bandwidth enhancement.

Keywords: CMOS; radio frequency; power amplifier; wireless communication; efficiency; linearity; bandwidth

Citation: Hamid, S.S.; Mariappan, S.; Rajendran, J.; Rawat, A.S.; Rhaffor, N.A.; Kumar, N.; Nathan, A.; Yarman, B.S. A State-of-the-Art Review on CMOS Radio Frequency Power Amplifiers for Wireless Communication Systems. *Micromachines* **2023**, *14*, 1551. https://doi.org/10.3390/mi14081551

Academic Editor: Zhiming Chen

Received: 19 June 2023
Revised: 28 July 2023
Accepted: 29 July 2023
Published: 1 August 2023

1. Introduction

Wireless communication systems based on radio frequency (RF) technology have become ubiquitous in electronic devices, particularly for mobile applications. As the number of subscribers continues to increase, the demand for RF communication systems also grows. Figure 1 illustrates the trend in mobile subscriptions by technology up to recent times [1]. The power amplifier (PA) block is a crucial component in transceiver design, as it consumes a significant amount of DC power. III-V compound semiconductors such as GaN HEMT, GaAs HBT, Si Bipolar, and SiGe HBT are commonly used to fabricate PAs, as they offer high linear output power and efficiency compared to the silicon CMOS process. However, these PAs are limited in their ability to integrate with CMOS baseband chips, hindering the development of a true single-chip radio or system-on-chip (SoC) [2]. To address this limitation, there was a recent focus on researching CMOS-based RF integrated circuits (RFICs) for wireless mobile applications [3–10]. However, CMOS-based PAs have several disadvantages due to their lower breakdown voltage, their higher substrate loss, and the unavailability of back via holes for the ground connection, which limit the achievable output power and efficiency compared to their III-V counterparts [11–13]. Despite their disadvantages, CMOS-based PAs offer a promising alternative due to their cost-effectiveness and potential for integration with other circuits. Several techniques were developed to improve the efficiency, linearity, and bandwidth performance of CMOS PAs [14–17].

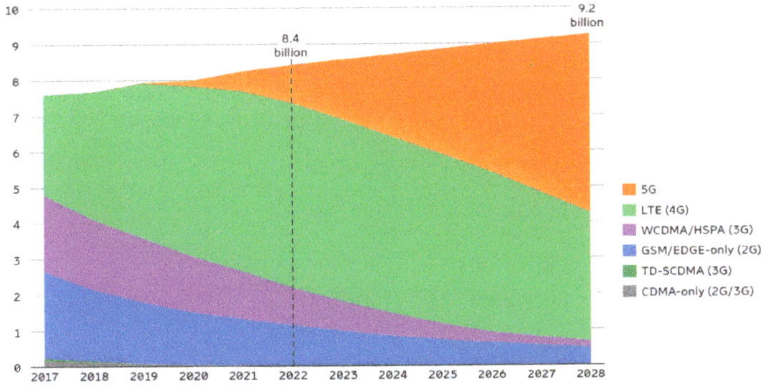

Figure 1. Mobile subscription trend [1].

Efficiency enhancement techniques are usually employed for highly linear PAs that consume more DC power. In contrast, linearity enhancement techniques are commonly employed for switching-based PAs, which have a low conduction angle that contributes to their non-linearity. Efficiency enhancement techniques target the efficiency of linear power amplifiers, while linearization techniques improve the linearity of highly efficient but non-linear power amplifiers [18]. As a result, there is an inevitable trade-off between efficiency and linearity performance in CMOS PAs. Recently, the bandwidth enhancement of CMOS PAs was also widely researched [19].

Wireless communication standards have rapidly expanded in recent years, with different applications utilizing various frequency bands such as L, S, C, UHF, and VHF, which are utilized by applications such as WLAN, Bluetooth, LTE, RFID, GSM, GPS, etc. [20–22]. To accommodate multiple standards in a single system, there is a growing need for CMOS PAs with enhanced bandwidth performance [23]. This state-of-the-art review paper provides an overview of the trending research efforts on the development of efficient and linear CMOS PAs with improved bandwidth performance for wireless mobile applications. This paper also discusses the limitations and challenges associated with CMOS PAs and outlines possible directions for future research.

2. Performance Metrics of CMOS Power Amplifier

Here, the terminologies and concepts associated with CMOS PAs' performance such as output power, efficiency, stability, intermodulation distortion (IMD), adjacent channel power ratio (ACPR), error vector magnitude (EVM), and amplitude-to-amplitude (AM–AM) and amplitude-to-phase (AM–PM) modulations are presented.

2.1. Output Power

Achieving optimal output power in PAs is crucial, and it is a vital parameter that varies depending on the wireless application requirements. For example, to effectively transmit data, cellular applications require a linear output power of 28 dBm [24], while Bluetooth demands a significantly lower output power of 0 dBm [25]. Additionally, the distance of the signal transmission between the transmitter and receiver is directly proportional to the output power level, meaning that a higher output power level results in an extended transmission distance.

An example of this concept is observed in Bluetooth technology, where a Class 1 specification necessitates a linear output power of 20 dBm, resulting in a signal transmission distance of 100 m. In contrast, a Class 3 specification only requires an output power of 0 dBm, limiting the signal transmission distance to 1 m [26]. To determine the output power of the PA, (1) can be utilized with reference to Figure 2, assuming an output load resistance of R.

$$P_{out} = \frac{\left(\frac{V_{p-p}}{2}\right)^2}{2R_L} \quad (1)$$

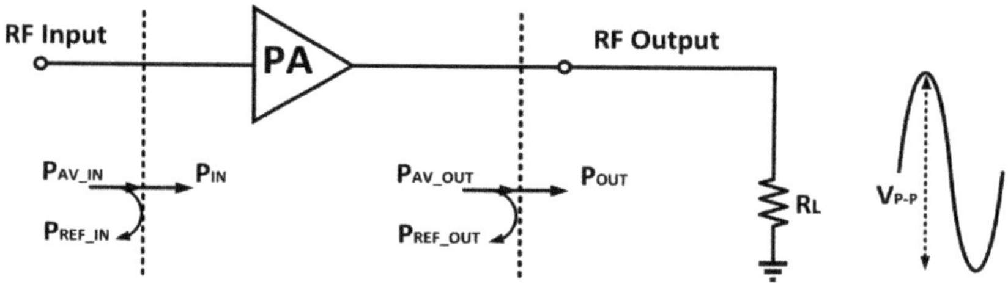

Figure 2. Depiction of an output power definition in a PA.

2.2. Efficiency

Efficiency is a crucial factor when it comes to assessing power consumption in PAs. Specifically, the drain efficiency and power added efficiency (PAE) are two commonly used parameters in the evaluation of PAs. In PAs, efficiency represents the amount of DC power that is converted into RF power. The PAE of a PA is a critical parameter that is typically used to assess the efficiency of PAs. In summary, drain efficiency and PAE play significant roles in evaluating the power consumption of PAs. The PAE of a PA is defined as [27]:

$$PAE = \frac{P_o - P_{in}}{P_{DC}} \quad (2)$$

$$\frac{P_o}{P_{DC}}\left(1 - \frac{1}{G}\right) = \eta_D\left(1 - \frac{1}{G}\right) \quad (3)$$

$$\eta_D = \frac{P_o}{P_{DC}} \quad (4)$$

where P_o is the output signal power, P_{in} is the input signal power, P_{DC} is the DC power consumption, G is the gain of the PA, and η_D is the drain efficiency of the PA. The drain efficiency and PAE are commonly close to each other when the gain of the PA is high.

2.3. Stability

In PAs, stability is a crucial parameter, particularly for high-gain PAs. It is worth noting that PAs that are stable under 50 Ω load terminations may not be stable in realistic wireless environments due to antenna impedance mismatches that occur, such as those caused by nearby objects [28–35]. Therefore, achieving an unconditionally stable characteristic is necessary for PAs to remain stable despite load impedance mismatches [36]. The stability of a PA can be determined from Rollett (K), which has to be larger than unity. A PA is unconditionally stable if the value of $K > 1$. Meanwhile, if $0 < K < 1$, it is considered conditionally stable, and, if $K < 0$, it is considered unstable. The K-factor is defined as

$$K = \frac{1 - |S_{11}|^2 - |S_{22}|^2 + |\Delta|^2}{2 \cdot |S_{11}||S_{22}|} \quad (5)$$

where

$$|\Delta| = |S_{11} \cdot S_{22} - S_{21} \cdot S_{12}|$$

S_{11} is the input port voltage reflection coefficient, S_{12} is the reverse voltage gain, S_{21} is the forward voltage gain, and S_{22} is the output port voltage reflection coefficient.

2.4. Intermodulation Distortion

Due to the non-linearity that exists in the PA, it produces numerous extra terms at innumerable frequencies. The transistors utilized in the PA produce non-linearity due to the non-linear current and non-linear capacitance that rely on the device voltages [37,38]. The main non-linear elements that effect the PA's performance are the transconductance (g_m), drain-source capacitance (C_{ds}), and gate-source capacitance (C_{gs}) [39]. The Taylor series expansion can be used in order to determine the polynomial transfer function of non-linearity. The significant terms are the first, second, and third orders, which represent the gain, squaring, and cubing functions [40]. The PA's non-linearity can be delineated by

$$V_o(t) = \dashv_o + \dashv_1 V_i(t) + \dashv_2 V_i^2(t) + \dashv_3 V_i^3(t) \tag{6}$$

Furthermore, the IMD that represents the linearity performance of a PA is determined by conducting a two-tone signals test in which approximately two spaced fundamental signal tones are supplied at the input of the PA. The amplitude of the two-tone signals is raised up to a certain level in which the third order products generate a signal with a higher amplitude above the noise floor [41–43]. The third order IMD (IMD3) product is the primary concern that needs to be dealt with because it is generated in-band of the fundamental signal frequency. The in-band-generated IMD product is difficult to eliminate using filters. Higher-order IMD products are small in amplitude to induce distortions and can be eliminated using filters [44]. The two-tone signal applied at the PA's input is given as

$$V_i(t) = \nu \cos(\omega_1 t) + \nu \cos(\omega_2 t) \tag{7}$$

Substituting (7) into (6) yields

$$\begin{aligned} V_o(t) = a_o + a_1\nu\,[\cos(\omega_1 t) + \cos(\omega_2 t)] + a_2\,\nu^2[\cos(\omega_1 t) + \cos(\omega_2 t)]^2 \\ + a_3\nu^3[\cos(\omega_1 t) + \cos(\omega_2 t)]^3 \end{aligned} \tag{8}$$

From (8), the generated IMD components up to third order are shown in Table 1.

Table 1. IMD components.

Order	Terms	a_1v^3	a_2v^2	a_3v^3
0	0		1	
1st	ω_1	1		9/4
	ω_2	1		9/4
2nd	$2\omega_1$		1/2	
	$2\omega_2$		1/2	
	$\omega_1 \pm \omega_2$		1	
3rd	$3\omega_1$			1/4
	$3\omega_2$			1/4
	$2\omega_1 \pm \omega_2$			3/4
	$2\omega_2 \pm \omega_1$			3/4

Thus, the amplitudes of both the upper and lower *IMD*3 products are defined as

$$IMD3 = \frac{3}{4}a_3\nu^3 \tag{9}$$

Figure 3 illustrates the output power spectrum of a PA subjected to IMD. It can be observed that the upper and lower *IMD*3 products are generated in-band approximately to the fundamental tones at $2\omega_2 - \omega_1$ and $2\omega_1 - \omega_2$, respectively.

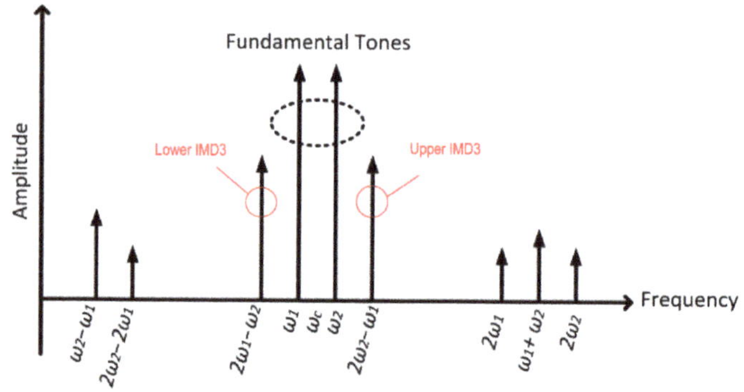

Figure 3. The IMD components induced by a PA.

2.5. Adjacent Channel Power Ratio

The adjacent channel power ratio (ACPR) is also a crucial metric that needs attention when a modulated signal is used as the input for the PA [45]. The ACPR values typically depend on the application of the wireless communication. Each wireless standard has adjacent channels located at different frequency offsets. For example, according to the 3GPP standard, the ACLR should be no more than −33 dB at the offsets of ±1.6 MHz and no more than −43 dB at the offsets of ±3.2 MHz. When a PA is operated with a modulated signal, it generates a bandwidth expansion that is commonly known as spectral regrowth. Spectral regrowth is a form of distortion since it results in adjacent channel interference [46,47]. Figure 4 is an exemplar of the output spectrum that illustrates the phenomena of spectral regrowth.

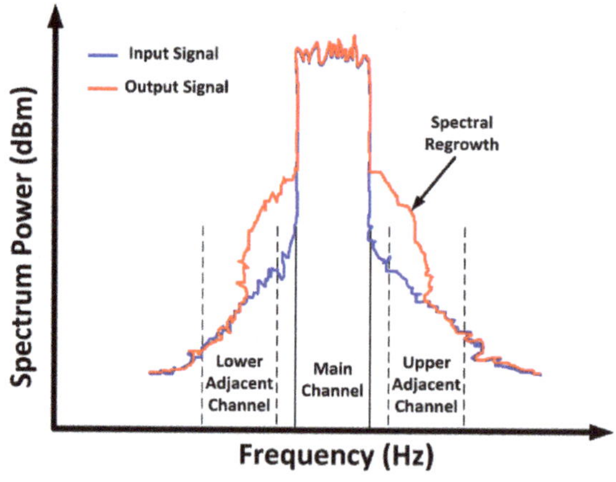

Figure 4. Spectral regrowth of a PA under a modulated signal.

The ACPR is defined in a frequency domain as the power ratio between the distortions in the sideband channels and the signal in the main channel. The ACPR is represented as follows:

$$ACPR_{adj} = 10 \log \frac{\int_{adj} S(f)df}{\int_{main} S(f)df} \tag{10}$$

where *S(f)* is the output signal's power spectral density (PSD), and the term *"adj"* can be either the first lower or upper channel.

2.6. Error Vector Magnitude

Another vital metric to determine the linearity of a PA is the error vector magnitude (EVM). The EVM is a time domain parameter that is represented in terms of a constellation diagram. The rotation and compression of the constellation points are the primary contributor to the constellation distortion, as delineated in Figure 5.

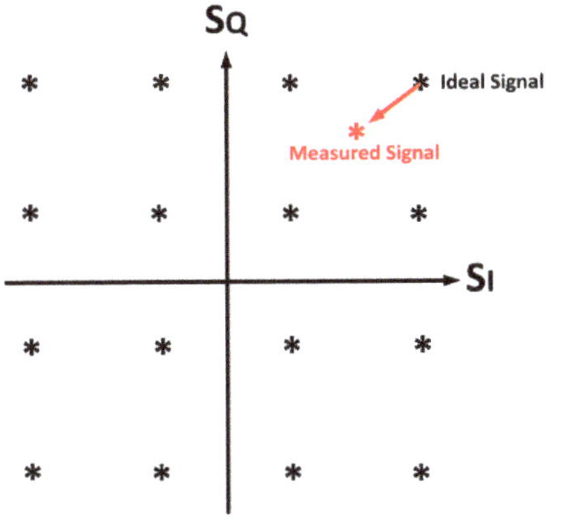

Figure 5. EVM constellation diagram.

The error vector is expressed as the vector discrepancy of the measured and ideal constellation points in the I/Q plane. The *EVM* is a customary figure of merit (FoM) utilized to interpret the distortions inside the operating band that deteriorate the bit error ratio (BER). The *EVM* is given as

$$EVM = \sqrt{\frac{\Sigma_{m=1}^{M}\left|S_{ideal,m} - S_{measured,m}\right|^2}{\Sigma_{m=1}^{M}\left|S_{ideal,m}\right|^2}} \quad (11)$$

2.7. Amplitude-to-Amplitude and Amplitude-to-Phase Modulations

The linearity of the PA can also be determined through the amplitude-to-amplitude (AM–AM) and amplitude-to-phase (AM–PM) modulations. When the PA is subjected to a large-amplitude input signal, the amplitude of the output signal tends to compress, and this decreases the linear output power that can be achieved. Moreover, the phase shift reflects the time delay in which, for an ideal case, it is undeviating at all amplitude levels. However, this is not the case in a real application, since the AM–PM modulation is not constant due to the existing non-linearity. A modulated input signal is expressed as [48]

$$\kappa\,(t) = A(t)\cdot\,\cos\left[\,\omega_c\,t + \theta(t)\right] \quad (12)$$

where $A(t)$ is the instantaneous amplitude, and $\theta(t)$ is the instantaneous phase.

AM–AM reflects the interrelation between the output signal's amplitude and $A(t)$. Meanwhile, AM–PM is the interrelation between the output signal's phase deviation and $A(t)$ [49]. Hence, the output signal is represented as

$$y(t) = g\,(A(t))\cdot\,\cos\left[\,\omega_c\,t + \theta(t) + \psi(A(t))\right] \quad (13)$$

where $g(A(t))$ and $\psi(A(t))$ are the AM–AM characteristics and AM–PM expression, respectively. The AM–AM and AM–PM characteristics are depicted in Figures 6 and 7, respectively.

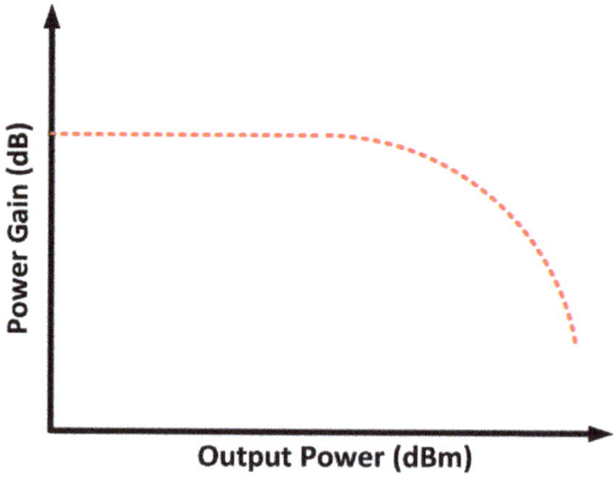

Figure 6. AM–AM feature of a PA across the output power.

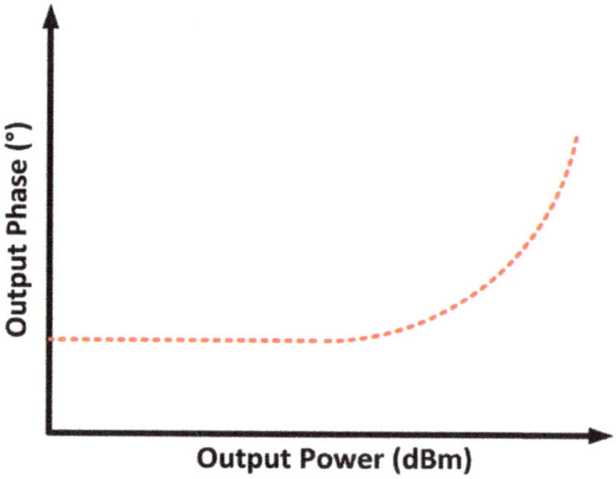

Figure 7. AM–PM feature of a PA across the output power.

3. Efficiency Enhancement Techniques

Several techniques were researched and implemented as efforts to enhance the efficiency of CMOS PAs. Commonly utilized efficiency enhancement techniques include the on-chip transformer, Doherty, envelope tracking (ET), envelope elimination and restoration (EER), and out-phasing PAs.

3.1. On-Chip Transformer Power Amplifier

Numerous efforts were applied to utilizing the configurations of on-chip transformers by integrating them together with CMOS PAs. On-chip transformers have various convenient features such as impedance matching, DC isolation, ESD protection, differential-to-single-ended conversion, and single-ended-to-differential conversion. On-chip transformer structures can be classified according to how they are connected. The structures include series combining (voltage mode) and parallel combining (current mode) [50]. The aforementioned structures are delineated in Figure 8.

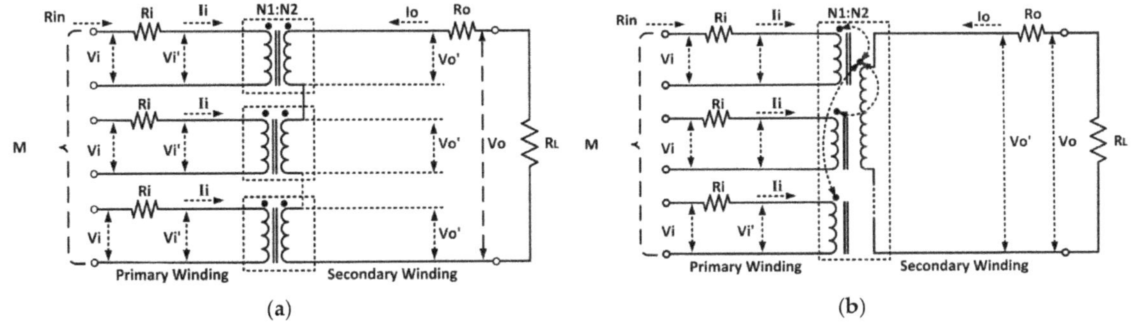

Figure 8. (**a**) Series transformer combination structure. (**b**) Parallel transformer combination structure.

As the name implies, the series combination transformer has its secondary coils connected in series. The series implementation exhibits a voltage addition on the secondary side, which yields higher output power. This is due to the equal current of V_O/R_L conducted by the winding of each transformer on the secondary side, as illustrated in Figure 8a [51]. The impedance observed by each amplifier is n times higher compared to the impedance seen when it is directly connected to R_L. This is advantageous for the driver amplifier design and for the mitigation of the parasitic effects on the primary side. The transformer has to be highly symmetrical because mismatches could occur, which degrades the maximum efficiency and output power that can be achieved by the PA [52].

By utilizing a series combination transformer, a CMOS PA achieving watt-level output power was proposed by Aoki et al. (2002) [53]. The CMOS PA is employed in a single chip with integrated input and output matching networks. A distributed active transformer (DAT) utilizing a series combining structure is employed for power combining and impedance matching, enhancing the efficiency of the PA [54]. However, the implementation consumes a huge active area on the chip. Meanwhile, a CMOS PA with a smaller footprint was achieved by Javidan et al. (2010) by combining a transformer with different sizes inside each other [55]. The proposed transformer is designed with multiple primary single-turn inductors and a single secondary multi-turn inductor.

The output of multiple PAs can be combined by using a series of combination 1:1 ratio transformers [56]. The output power of the PA can be varied by switching the individual transformers on or off [57]. A. Afsahi et al. (2010) utilized a transformer in order to parallelly combine signals to reduce losses on the secondary side as well as to achieve a better signal symmetry [58]. However, the number of turns in the secondary winding increased, which degrades the self-resonance frequency as well as consumes a huge chip area.

Moreover, K.H An et al. (2007) proposed a monolithic voltage-boosting parallel primary transformer that was utilized for power combining [59]. The transformer gave a voltage-boosting effect by increasing its turn ratio between the primary and secondary windings. An increased current was established in the secondary winding by interweaving multiple primary windings in parallel. An efficient power combining method was achieved by elevating both the secondary voltage and current. An interleaved transformer structure was proposed by A.R. Belabad et al. (2013), which achieved better efficiency compared to a 1:1 ratio transformer [60]. Although the interleaved configuration reduced the Q-factor and self-inductance, it had an enhanced coupling factor that is beneficial for transferring current via its magnetic field. The proposed transformer had two input ports and an output port. The drain terminals of different PAs were connected to the input ports of the transformer while the load was connected to its output port.

Furthermore, H. Ahn et al. (2017) proposed a dual-mode autotransformer-based parallel combining transformer (ABPCT) that provides high efficiency [61]. An individual PA was connected on the secondary side while multiple individual PAs were connected on the primary side. For high-output power operation, all the PAs were turned on. Contrarily, for low-output power operation, one or two discrete PAs were specifically turned on. The high- and low-power modes were selected by controlling the gate bias of each PAs. Autotransformers have lower power losses due to their reduced total series resistance, which is contributed to by the primary winding that is also a part of the secondary winding. An autotransformer-based PA attained an efficiency of 34% in low-power mode and 38% in high-power mode.

In addition, J. Tsai et al. (2019) proposed a CMOS PA with a transformer-based two-stage dual-radial power splitting and combining mechanism [62]. Several advantages of this proposed transformer included in-phase signal splitting and combining, a compact network, the uniform distribution of DC supplies, and a symmetric supply and return path of the DC current. By utilizing the multi-layer structure of CMOS technology, the radial power splitting plane was implemented on a lower metal layer below the radial power combining plane. This contributed to a compact chip size since both the power splitting and combining shared the same chip area on top of each other. A shielding structure was employed to enhance the isolation between the input redial power splitting and output radial power combining networks. The maximum output power delivered by the PA was 29 to 31 dBm, with a maximum efficiency of 15–25%. J. Tsui (2019) also proposed a 5.3 GHz CMOS PA with a folded radial power splitting and binary power combining on-chip transformer [63]. The proposed architecture of the transformer provided an in-phase signal splitting and combining mechanism. The transformer was also designed with a compact size with uniform DC distribution as well as a symmetric DC current path. At the operating frequency of 5.3 GHz, the CMOS PA achieved a maximum output power of 31.3 dBm, with a peak PAE of 22%.

Another on-chip transformer-based CMOS PA was presented by H. Choi et al. (2020), in which a 1:1 turn spiral transformer balun was implemented with a stacked main PA configuration in order to sustain a high voltage supply [64]. The main PA was constructed in a differential push–pull configuration in order to attain high output power and stability. The 1:1 transformer was employed with an equal coupling line length and an equal inductance in order to achieve an optimum impedance with minimum losses. The proposed CMOS PA also utilized an anti-phase biasing technique to achieve a higher linear output power. The CMOS PA achieved a maximum output power of 29.9 to 30.3 dBm, with 40% to 45% across its operating bandwidth of 0.82 to 1 GHz.

T. Wang et al. (2021) presented an on-chip transformer-based digital CMOS PA in which the transformer acted as a balance-compensated matching network [65]. A series-loaded compensation capacitor was utilized to enhance the balance response of the transformer, which improved the efficiency of the CMOS PA. The transformer was implemented in a four-way series combining configuration. The series combining configuration aided in reducing the voltage stress on the transistors and the impedance transformation ratio of the matching network. The designed transformer was capable of simultaneously providing four-way power combining, balance compensation, wideband impedance matching, and differential-to-single-ended conversion. The proposed CMOS PA delivered a maximum output power of 32.4 dBm at 2 GHz. The maximum drain efficiency achieved was 53.8% at 1.8 GHz.

B. Yang et al. (2021) presented a CMOS PA with a reconfigurable self-coupling cancelling transformer [66]. The proposed transformer had an improved tuning range of the turn ratio that was utilized to boost the PA's efficiency at deep back-off output power. The core of the PA was constructed in a quadrature digital PA configuration with I-Q cell sharing and a hybrid Doherty architecture. The PA also employed a switched/floated capacitor for impedance tuning. The CMOS PA achieved a maximum output power of 30.3 dBm, with 36.6%, the highest system efficiency, at 2.4 GHz. B. Yang et al. (2023) also proposed a watt-level CMOS digital PA using a reconfigurable power-combining transformer [67]. The transformer as employed in a voltage-mode power combining configuration in order to attain high output power across a wide frequency bandwidth. The output power of the CMOS PA at the fundamental signal was further enhanced by using a LC circuit that suppresses the harmonics. The reconfigurable output matching network was constructed using the power-combining transformer together with four switched capacitors at the inputs and a fixed capacitor and output. Additionally, the dynamic range of the digital PA was enhanced by mitigating the LO leakage by using a 12-bit power digital-to-analog converter (power DAC). The CMOS PA delivered a maximum output power of 32.67 dBm, with a peak PAE of 35.5% at 2 GHz.

3.2. Doherty Power Amplifier

The Doherty PA (DPA) was initially proposed by W. H. Doherty, comprising a carrier and a peaking amplifier [68]. A typical block diagram of a classic DPA is depicted in Figure 9 [69]. The back-off efficiency of the PA is improved through its load impedance variation. This is realized by implementing a current source at the output terminal. Figure 10 illustrates the concept of the DPA.

Based on Figure 10, the auxiliary amplifier is denoted by "Aux", and the main amplifier is denoted by "Main" [70]. The auxiliary amplifier is commonly operated in class C mode while the main amplifier is operated in class AB mode. When the auxiliary amplifier is inactive, the main amplifier faces a load resistance of R_L. Whereas, if the auxiliary amplifier is active and generates a current, I_{aux}, the main amplifier, sees a load impedance as

$$Z_{main} = R_{LOAD}\left(1 + \frac{I_{\text{aux}}}{I_{main}}\right) \tag{14}$$

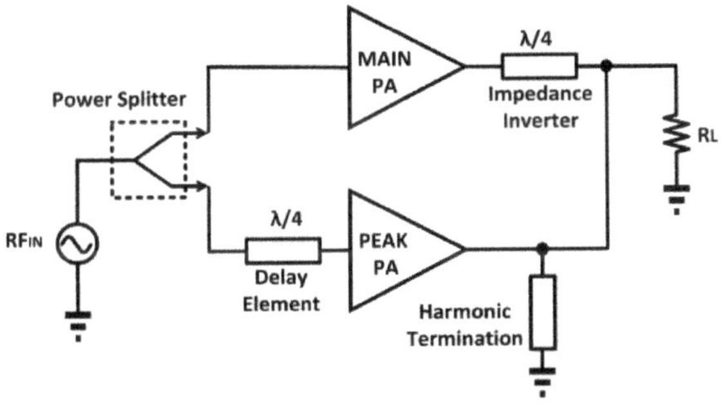

Figure 9. Block diagram of a classic DPA.

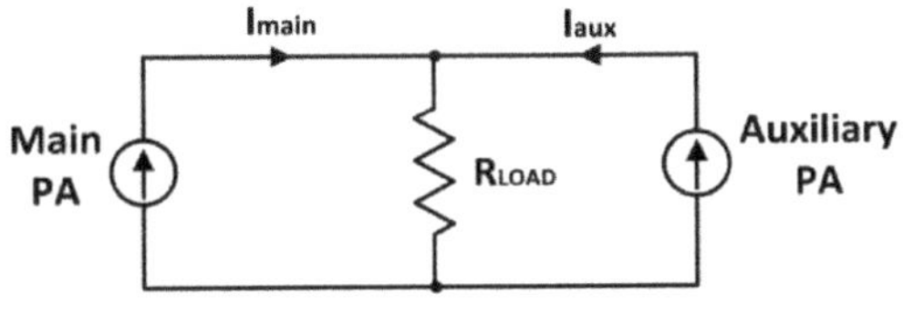

Figure 10. The concept of Doherty PA.

From (14), it is evident that the load impedance of the main amplifier can be varied by controlling the source current of the auxiliary amplifier (I_{aux}). This feature of the DPA is utilized to enhance its back-off efficiency, as illustrated in Figure 11 [71]. A constant efficiency across output power was attained by utilizing the DPA technique in CMOS, as presented by N. Deltimple et al. (2015) [72]. S. Hu et al. (2016) proposed a class G DPA that enhances the efficiency at deep back-off output power [73]. A hybrid combo of class G and Doherty improved the deep-back off efficiency without extra complexity at the input and output passive networks.

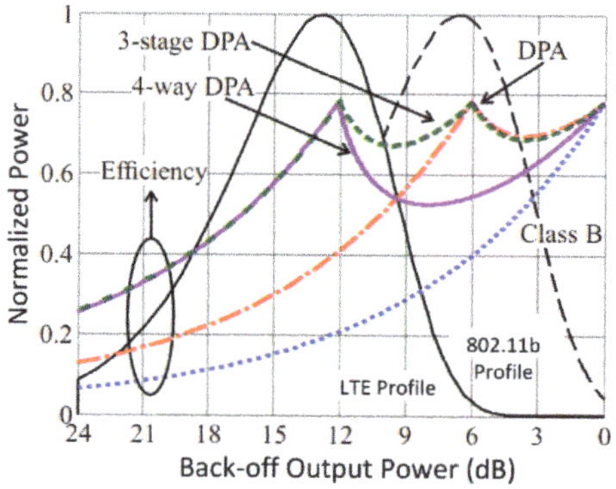

Figure 11. The efficiency profile of DPA.

E. Kaymaksut et al. (2015) presented a CMOS dual-mode Doherty PA with a four-way symmetrical hybrid transformer [74]. The transformer mitigated the impedance imbalance that exists in the PA. The PA employs a switching mode technique with the combination of the Doherty technique, which realizes the high-power and low-power modes. The PA also utilized an adaptive bias circuit that further enhances its linearity and back-off efficiency across its frequency of operation. The proposed transformer utilized the combination of a distributed active transformer (DAT) and a figure-eight-shaped transformer [75,76]. The designed transformer was a 2×2 hybrid transformer that utilizes the combination of two DATs with a figure-eight-shaped transformer. In LP mode, only two sub-PAs are switched on, while in HP mode, all the PAs are made active. The PA operates from 1.7 to 2.1 GHz, with a maximum output power of 26–28.2 dBm, as well as a peak efficiency of 25.5–34%. The achieved linear output power was 21–23.4 dBm, with a corresponding efficiency of 16–23.4%.

Further, C.S. Levy et al. (2016) presented a series Doherty PA (SDPA) that is capable of delivering a higher output power than a conventional DPA [77]. In an SDPA, the load voltage is two times higher compared to the voltage of each amplifier separately because of the parallel load impedance of the peaking amplifier. To achieve an efficiency approximate to that of class B operation, the main amplifier is operated in deep class AB operation. In addition, S. Traiche et al. (2017) presented a highly efficient fully integrated CMOS DPA [78]. The proposed DPA comprised a main and an auxiliary amplifier that are operated in class AB and class B, respectively. Both amplifiers utilized the cascode structure to endure the high supply voltage and to achieve high input–output isolation. A Wilkinson divider was used at the input for signal splitting. As for the transmission line, a π-network-based dephasing circuit and impedance inverter were utilized, as well as resonant tanks, to terminate the harmonics.

In addition, D. Jung et al. (2019) presented a Doherty CMOS PA with multi-gated transistors (MGTRs) in the main PA core [79]. This Doherty CMOS PA also employed a class C mode auxiliary PA and a synthesized-transformer-based parallel output network with a second-harmonic control circuit. Each MGTR consisted of four transistors with different sizes and was biased to attain a near-zero third-order transconductance operation. The MGTR and the second-harmonic control circuits both aid in suppressing the third-order intermodulation distortion of the proposed Doherty PA. The MGTR Doherty CMOS PA achieved a maximum output power of 27.2 dBm, with 24.5% PAE at 5.8 GHz.

Y. Yin et al. (2020) proposed a CMOS Doherty PA using digital implementation with a parallel combining transformer for deep back-off efficiency enhancement [80]. The parallel combining transformer offers a dynamic load modulation for the digital Doherty PA in an ultra-compact size. The designed digital Doherty PA consisted of four sub-PAs that are controlled in a precise manner. The load modulation was performed by connecting each of the differential sub-PA pairs to the terminals of the two adjacent primary coils of the power combiner. The digital Doherty PA delivered a maximum output power of 21.4 dBm, with a peak PAE of 31.3% at 1.5 GHz.

3.3. Envelope Tracking Power Amplifier

In an envelope tracking power amplifier (ETPA), a supply modulation mechanism is employed to manage the supply voltage of the PA with the corresponding RF input voltage while maintaining a fixed load resistance [81]. This contributes to the efficiency improvement at the low-output power region by minimizing the power dissipation for systems with a high PAPR [82]. The ET supply modulator tunes the supply voltage level of the PA with respect to the time-varying envelope of the RF input signal in order to keep the PA in stable compression mode. This contributes to the efficiency enhancement of the PA. Figure 12 illustrates the voltage difference between the supply voltage and the RF output signal under a modulated signal [83].

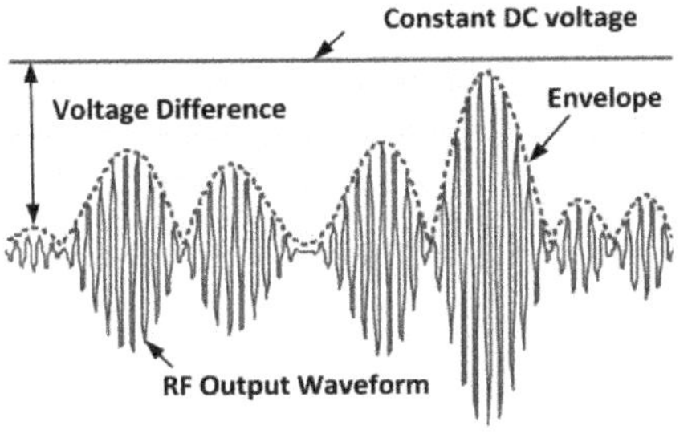

Figure 12. Voltage difference between supply voltage and modulated RF output signal waveform.

An envelope amplifier is utilized for biasing the RFPA according to the input envelope RF signal, as depicted in Figure 13. An envelope detector on the input path in employed to retrieve the envelope information. The envelope information can also be retrieved by utilizing digital baseband processing [84].

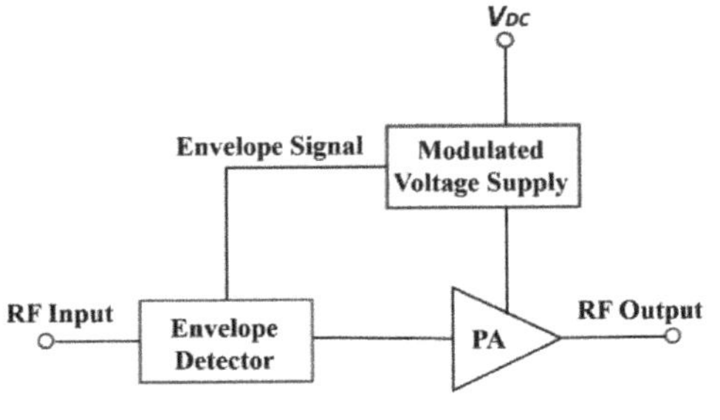

Figure 13. Block diagram of the ETPA.

Figure 14 depicts the block diagram of an ETPA together with its relative waveforms. The ET method is commonly integrated with a linear power amplifier (LPA), and the RF signal amplified contains both the amplitude and phase information. The power supply in an ETPA is mainly focused on enhancing the efficiency of the PA, and its overall efficiency includes both the efficiency of the PA and the envelope amplifier.

J. L. Woo et al. (2014) introduced a dynamic stacked CMOS ETPA that achieved high efficiency during low-output power operation [85]. The proposed ETPA achieved high efficiency by utilizing a 3.4 V maximum voltage. Furthermore, B. Park et al. (2016) proposed an ETPA with an improved ET supply modulator [86]. By integrating this supply modulator, an enhanced efficiency was achieved at all output power levels. There were also second-order harmonic control mechanisms employed at the input and output of the ETPA that further improved its efficiency. Moreover, an ETPA with a dual-mode supply modulator was presented by Ham et al. (2016) [87]. A dual-mode supply modulator was utilized for envelope tracking and average power tracking. This modulator was designed based on a hybrid buck converter comprising switching and wideband linear amplifiers. This ETPA achieved a peak efficiency of 45.4% under the envelope tracking mode.

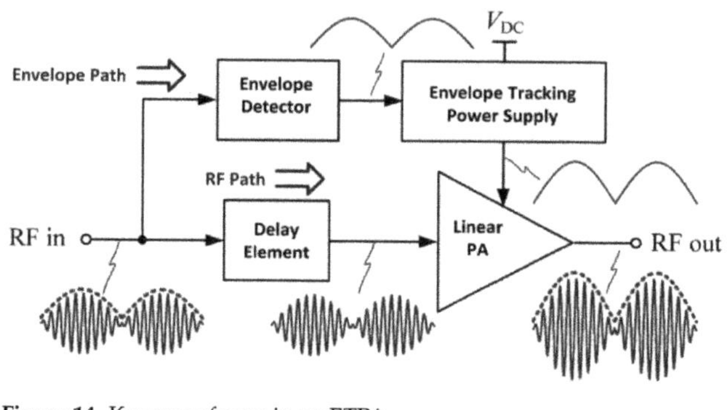

Figure 14. Key waveforms in an ETPA.

3.4. Envelope Elimination and Restoration Power Amplifier

The envelope elimination and restoration (EER) technique was introduced by L. R. Khan in 1952 [88]. Contrary to ETPA, the EER technique is implemented in a non-linear PA (NLPA). EER adopts a supply modulation method in which it splits the RF modulated signal into envelope and phase signals and later combines it at the output of the NLPA. A high-speed supply modulator is employed for the supply voltage modulation of the NLPA. Figure 15 delineates the block diagram of an EERPA together with its key waveforms.

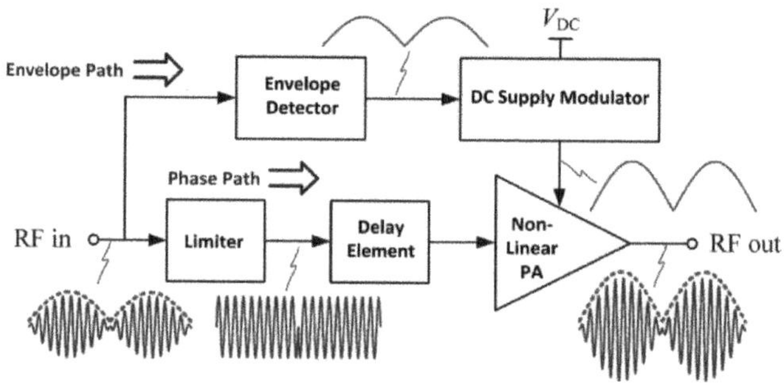

Figure 15. Key waveforms in an EERPA.

Referring to Figure 15, the RF input signal that consists of amplitude and phase information is handled by two different paths, which are the envelope and phase paths. At the envelope path, the input signal's envelope is identified by the envelope detector and is utilized as the reference voltage for the supply modulator. The supply modulator then generates an output voltage that follows the input signal's envelope and supplies the voltage to the NLPA. In contrast, at the phase path, the limiter is utilized to eliminate the input signal's envelope information and generate a constant amplitude signal with only the phase information. The amplitude information is restored by the supply voltage given by the supply modulator. The delay unit is utilized to synchronize both the envelope and phase paths.

A. Mamdouh et al. (2020) presented a fast transient supply modulator for an EERPA [89]. The efficiency was improved at a high frequency due to the fast transient response of the supply modulator compared to conventional hybrid mode supply modulators. K. Oishi et al. (2014) proposed an EERPA with an envelope/phase generator based on a mixer and a limiter that generates a wide-bandwidth phase signal [90]. A delay locked loop-based timing aligner was implemented with a variable high-pass filter to compensate for the timing mismatch. Furthermore, W. Yuan et al. (2015) explored a digital-based implementation and proposed an EERPA with a class E configuration and a digitally controlled current DAC modulator [91]. A switched capacitor-based DAC was presented to control an open-loop transconductor that serves as a current modulator. The current modulator modulated the amplitude of the current supplied to the class E configured PA.

X. Liu et al. (2019) proposed an EER envelope-shaping PA with an AC-coupling supply modulator [92]. The supply modulator comprised a 25 MHz three-level switching amplifier and a wideband-assisting linear amplifier. The AC-coupling method utilized the shaped envelope to mitigate the power consumption of the linear amplifier and generate a higher voltage than its supply. A high-efficiency, small-output voltage ripple and a fast transient response were achieved by implementing a switching amplifier with three-level real-time flying capacitor calibration. A multi-loop control was utilized in order to further increase the switching amplifier speed and, thus, enhance the overall efficiency. As for the PA, a mode-switching PA was implemented in a differential cascode structure, and bond wires were utilized as its DC-feed inductor. To improve its efficiency, the common-gate voltage of the cascode PA was dynamically biased with respect to the instantaneous supply voltage. A capacitive-charging-acceleration method was also implemented to further increase the efficiency [93]. The supply modulator achieved an efficiency of 88.7%, while the overall EERPA achieved an efficiency of 35.7%.

3.5. Out-Phasing Power Amplifier

An out-phasing PA utilizes the method of separately splitting and amplifying the signals then combining them at the output. The signal with amplitude modulation is given as the vector sum of two constant envelope phase-modulated signals. An improved efficiency is achieved by modulating the differential phase while keeping the amplitude of the input signal constant. For out-phasing techniques utilizing non-linear components, amplitude modulation is conducted by combining the outputs of two PAs THAT are supplied with constant envelope phase-modulated signals [94–98]. The architecture of the out-phasing technique is shown in Figure 16, and its principle of operation is depicted in Figure 17.

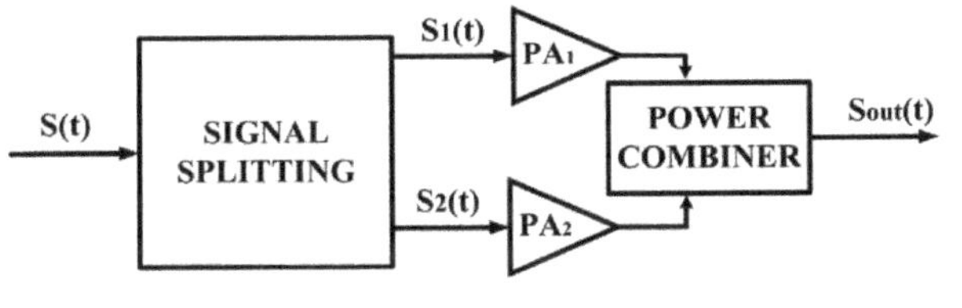

Figure 16. Architecture block diagram of out-phasing.

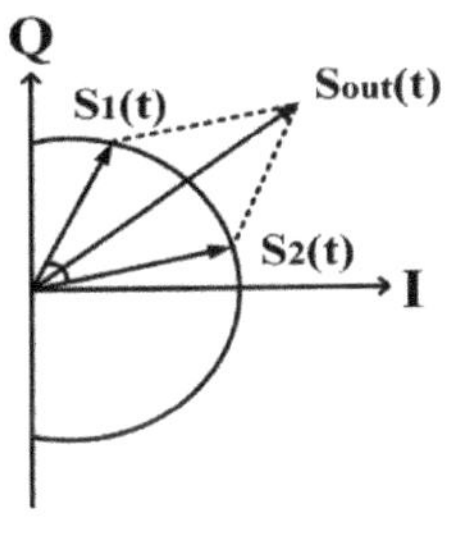

Figure 17. Operation principle of out-phasing.

A modulated signal can be represented, as shown in (15), and its amplitude and phase can be split into two constant amplitude and phase-modulated signals ($S_1(t)$ and $S_2(t)$), as shown in (16) and (17).

$$S_{out}(t) = A(t)\cos(\omega_c t + \varnothing(t)) \quad (15)$$

$$S_1(t) = \frac{A_{max}}{2}\cos(\omega_c t + \varnothing(t) + \theta(t)) \quad (16)$$

$$S_2(t) = \frac{A_{max}}{2}\cos(\omega_c t + \varnothing(t) - \theta(t)) \quad (17)$$

where $S_{out}(t) = S_1(t) + S_2(t)$, $A(t)$ is the time-dependent amplitude, and $\phi(t)$ is the time-dependent phase of the original signal, $S(t)$; ω_c is the angular frequency of the carrier, and A_{max} is the maximum value of $A(t)$. The angle $\theta(t)$ is defined as

$$\theta(t) = \cos^{-1}\frac{A(t)}{A_{max}} \quad (18)$$

Since $S_1(t)$ and $S_2(t)$ are constant envelope phase-modulated signals, the signal amplification can be conducted by utilizing a non-linear PA. The outputs of the PA can be joined by using a combiner circuit in order to obtain $S_{out}(t)$.

Z. Hu et al. (2016) proposed an RFDAC-based out-phasing PA with an integrated combiner [99]. Increased efficiency was achieved by utilizing a dynamic amplitude control in a quasi-load insensitive class E configured PA. The amplitude was controlled through a digitally divided method for the branch amplifier. Highly reactive and severe load modulation conditions were avoided through this method, which contributed to the minimization of power losses.

Furthermore, a CMOS class D configured multi-level out-phasing transmitter with on/off control logic method was presented by M. Martelius et al. (2016). The PA consisted of eight individual amplifiers with cascoded output stages [100]. The PAs were selectively activated in pairs to achieve different amplitude levels as a means of improving the back-off efficiency. A. Banerjee et al. (2017) proposed a CMOS class E configured out-phasing PA with a novel passive combining circuit [101]. The designed passive combining circuit contributed to achieving enhanced output power and efficiency. The combiner comprised a power enhancement circuit and an efficiency enhancement circuit integrated together. Moreover, A. Ghahremani et al. (2018) presented a class E out-phasing PA by conducting load-pull analyses [102]. These load-pull analyses were carried out to rotate and shift the power and efficiency contours. This method was utilized to improve the efficiency at deep back-off output power. The rotation of the power and efficiency contours were realized by tuning the LC tank's resonance frequency and duty cycle scaling factor. This class E PA consisted of transistors that serve as a square-wave input-signal-driven switch with a duty cycle of 50%. Parallel bond-wires with a high Q-factor were utilized as the DC-feed inductors for each PA. The cascade inverters employed in the PA were used as a driver for the switch. The duty cycle of the switch was controlled by using the off-chip control voltage. The switch capacitor banks were used to vary the resonance frequency of the LC tank at the switching nodes.

M. Martelius et al. (2020) proposed an out-phasing CMOS PA with the combination of a polar mechanism that implements a tri-phasing modulation [103]. The combination of the polar and out-phasing components mitigated the linearity-degrading effects while maintaining the back-off efficiency. The CMOS PA comprised eight class D units on a single chip. The power combiner of this CMOS PA was implemented using Marchand baluns that consist of input transmission lines and coupled-line architecture on a printed circuit board. This CMOS PA achieved a maximum output power of 29.7 dBm, with a peak PAE of 34.7% at 1.7 GHz.

A. Banerjee et al. (2020) presented a multi-mode class E out-phasing CMOS PA with an integrated passive combining circuit [104]. Multiple class E PAs were turned ON and OFF in order to configure the efficiency of the PA at low power levels. This out-phasing CMOS PA also utilized an efficiency enhancement circuit (EEC) which further assisted in improving the efficiency at back-off output power. In an out-phasing PA, the in-phase components are added together and increase the load power, while the out-of-phase components cancel each other out and reduce the load power. In order to enhance the back-off efficiency, the power loss due to the out-of-phase components should be minimized without trading off the in-phase operation. Thus, an EEC is employed for this purpose. The proposed EEC is realized using an inductor that is connected between the drains of the main PAs. Table 2 summarizes the recent state-of-the-art CMOS PAs with efficiency enhancement techniques.

Table 2. Summary of the recent state-of-the-art CMOS PAs with efficiency enhancement techniques.

Architecture	Year	Tech (nm)	V_{DD} (V)	Freq. (GHz)	Max. P_{out} (dBm)	Peak PAE (%)	Ref.
On-Chip Transformer	2010	180	3.3	0.9	29.5	24.0	[55]
	2010	65	3.3	2.4	33.5	37.6	[58]
	2013	180	3.3	2.4	32.5	37.9	[60]
	2017	40	2.8	1.75	24.7	38.1	[61]
	2019	180	5.0	5.2	30.1	22.2	[62]
	2019	180	3.3	5.3	31.3	22.0	[63]
	2020	180	4.6	0.82–1.0	30.3	45.0	[64]
	2021	40	2.5	1.2–2.8	32.4	53.8	[65]
	2021	40	2.4	2.4	30.3	36.6	[66]
	2023	40	2.5	2.0	32.67	35.5	[67]
Doherty	2015	65	2.5	2.5	23.4	24.7	[72]
	2016	65	3.0	3.71	26.7	40.2 *	[73]
	2015	40	1.5	1.7–2.1	28.2	34.0	[74]
	2016	45	2.4	14.0	22.0	24.0	[77]
	2017	130	2.6	3.0	26.0	48.0	[78]
	2019	55	5.5	5.8	27.2	24.5	[79]
	2020	40	1.1	1.5	21.4	31.3	[80]
ETPA	2014	320	3.4	0.837	25.9	47.5	[85]
	2016	180	4.7	1.7	28.5	36.6	[86]
	2016	180	3.3	0.78	24.0	45.4	[87]
EER	2020	130	3.0	1.0	26.5	79	[89] +
	2014	90	3.3	1.95	28.0	39	[90]
	2015	130	2.5	2.4	22.5	23.6	[91]
	2019	65	2.4	2.4	23.9	35.7	[92]
Out-phasing	2016	40	1.2	5.9	22.2	49.2 *	[99]
	2016	28	3.6	1.8	32.4	34.1	[100]
	2017	45	2.4	2.4	29.5	46.8 *	[101]
	2018	65	1.25	1.8	20.1	60.7	[102]
	2020	28	3.6	1.7	29.7	34.7	[103]
	2020	45	2.4	2.4	31.6	49.2 *	[104]

* drain efficiency; + simulated.

Figure 18 delineates the plot of the PAE achieved by the aforementioned techniques across the achieved output power. The proposed techniques are capable of delivering a maximum output power of more than 20 dBm. It can be observed from Figure 18 that the on-chip transformer and the out-phasing techniques are capable of delivering watt-level maximum output power for CMOS PAs. This is due to the usage of power-combining architectures in these techniques that boost the maximum output power. The PAE achieved by these techniques was mostly more than 30%. The highest PAE was achieved by using the EER technique; however, the research was limited to only simulated data.

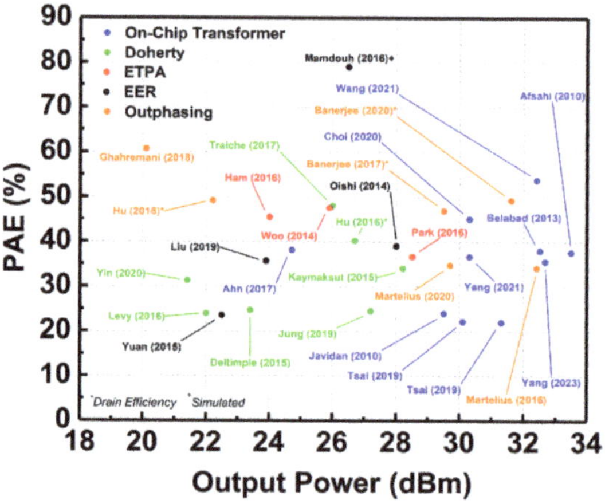

Figure 18. PAE across output power for the proposed efficiency enhancement techniques [55,58,60–67,72–74,77–80,85–87,89–92,99–104].

4. Linearity Enhancement Techniques

Non-linearity in PAs is mainly contributed to by signal distortions. In CMOS MOSFETs, the distortions are generated by the inherent substrate non-linearity, trans-conductance, and parasitic capacitances, mainly the gate-source capacitance (C_{gs}) [105–107]. AM–AM and AM–PM are some of the distortions that are caused by parasitic capacitances. The AM–AM distortion is minimal at a low frequency, and it increases as the frequency increases due to the non-linear C_{gs} [108].

AM–PM is mainly impacted by the MOSFET's transconductance, which can be mitigated with suitable input matching network phase compensation techniques [109]. Intermodulation distortion (IMD), especially the third-order IMD, is another type of distortion that impacts the linearity performance of PAs. In a two-tone test, the third-order IMD occurred close to the fundamental tones and was arduous to filter out. The third-order IMD increased by 3 dB for every 1 dB increase in the fundamental tones [110]. In addition, spectral regrowth is another form of distortion in the adjacent frequency channels in which the power regrows. These adjacent channels might be assigned for different communications, which generates perturbations [111].

High linearity is necessary to fulfill the stringent protocols. Hence, linearization techniques are utilized to mitigate the distortions inside the operating bandwidth and to reduce the interferences from the adjacent frequency bands. Numerous techniques were employed to enhance the linearity of CMOS PAs, which include feedback techniques such as Cartesian loop feedback, polar loop feedback, feedforward, linear amplification using non-linear components (LINC), analog pre-distortion (APD), digital pre-distortion (DPD), and adaptive biasing.

4.1. Feedback Technique

The feedback method is conducted to improve linearity by connecting a fragment of the output signal back to the input of the PA [112]. The most common feedback techniques are the Cartesian and polar feedbacks. The feedback technique is advantageous due to its simplicity. However, its input and output signals need to be simultaneously operated, which is impossible to be realized due to the delays in the circuits. This serves as the major bottleneck of the feedback method. Figure 19 delineates the feedback principle in which part of the output signal (feedback signal) is connected back into the PA's input after passing through a delay compensation mechanism, β.

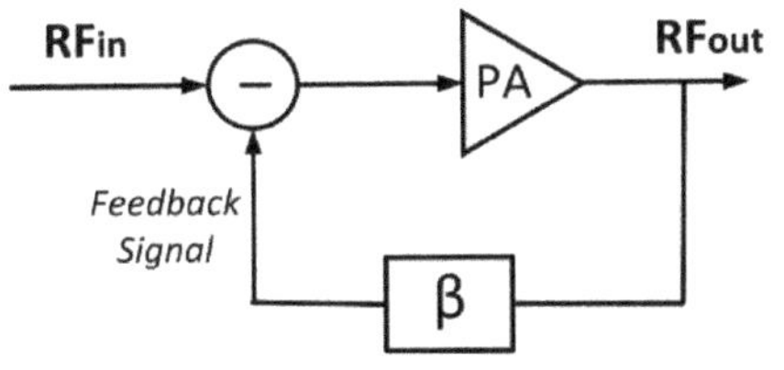

Figure 19. Principle of feedback technique.

4.1.1. Cartesian Feedback Technique

Cartesian feedback is a technique that encompasses the linearization of the overall transmitter system. In the Cartesian feedback method, the baseband I and Q signals are upconverted to the carrier frequency, and the signals are then amplified. The signals are later sampled and downconverted back into the quadrature component, which is connected back to the input of the transmitter.

Error amplifiers are employed to compare the signal to the original baseband input signals [113]. The delay that exists in the feedback loop limits the bandwidth of the transmitter. Figure 20 depicts a simple block diagram of a Cartesian feedback implemented in a PA [114].

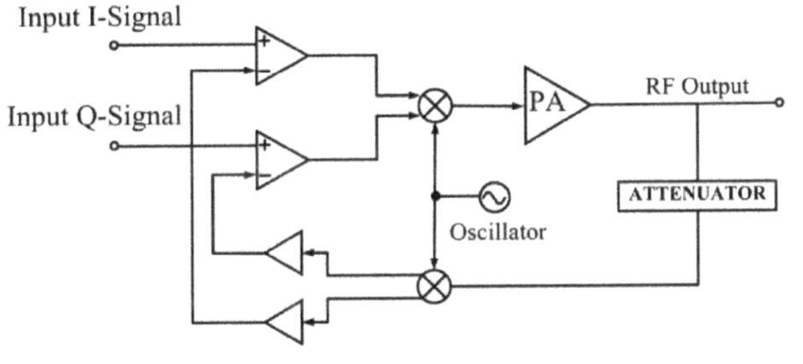

Figure 20. Cartesian feedback technique.

L. Tee et al. (2006) presented a Cartesian feedback implemented in a CMOS PA for EDGE application [115]. The PA comprised a pre-driver, a driver, and a main amplifier. The main amplifier was operated in class C, which attained high efficiency. A class AB transistor was integrated with the class C amplifier to achieve stability at all input levels at the output stage. Furthermore, a Cartesian feedback linearization for CMOS Zigbee PA was proposed by A. Atress et al. (2016) [116]. A class E operation was adopted to obtain a high efficiency. The PA was tested with $\pi/4$ DQPSK modulation at 868 MHz in order to verify its linearity performance. A resistor was used at the input matching network to improve the stability of the PA. An inductor and a capacitor connected in series in the output matching network were utilized for waveform shaping. The employed Cartesian feedback system was capable of lowering the spectral regrowth at the adjacent channels.

4.1.2. Polar Feedback Technique

Unlike a Cartesian feedback, which feedbacks the I and Q data, the polar technique feedbacks the amplitude and phase into the input of the system. The difficulty with this approach is that the feedback bandwidths of the amplitude and phase components are different from each other. This deteriorates the available loop gain to either the amplitude or phase path because one path needs a feedback bandwidth that limits the loop gain, while the other path requires a larger loop gain. Figure 21 shows the block diagram of a polar feedback system that addresses both amplitude and phase distortion with separated loops [117].

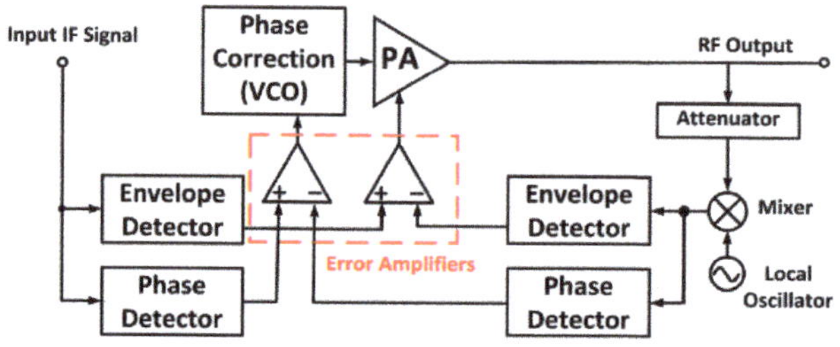

Figure 21. Block diagram of a polar feedback system.

In the polar feedback technique, a local oscillator and a mixer are employed to attenuate and downconvert the output signal of the PA. The downconverted output signal is compared with the envelope of the incoming intermediate frequency (IF). Moreover, the error amplifiers in the negative feedback loop are employed to control the bias condition of the PA.

A polar transmitter with a current-mode class D CMOS PA was presented by T. Nakatani et al. (2013). This polar transmitter also consisted of a digital pulse width modulation (DPWM) algorithm that was employed to mitigate spurious signals linked with the digital input signal envelope [118]. DPMW was mainly employed to minimize the noises induced by the clock via a minimum quantization error dithering (MQED) method. This approach spreads out clock spurs, reduces quantization errors, and mitigates peak spurs.

Further, S. Zheng et al. (2013) presented a digital polar transmitter comprising a six-bit CMOS PA array [119]. The proposed design employed a highly linear charge-mode switched-capacitor digital polar modulator (DPM) in place of the conventional switched-current DPM. A cascode common-source configuration in the differential structure was utilized in the PA. The replica of the main PA was integrated with a comparator and a resistor. The replica was employed to regulate the main PA's bias level with an analog feedback loop in which it mitigated the AM path's distortion.

4.2. Feed-Forward Technique

Feed-forward linearization is employed to mitigate distortions at the output of the PA. The feed-forward technique focuses on the subtraction of the harmonics and intermodulation components from the output spectrum of the PA [120]. Feed-forward comprises two cancellation loops based on two amplifiers: the main amplifier and the error amplifier. Figure 22 shows the block diagram of a feed-forward-based PA. The input signal is split into two, in which one portion is amplified via the main PA, and another portion is processed via the delay element. The feed-forward method is advantageous for wideband linearization, but it increases the implementation cost and circuit complexity.

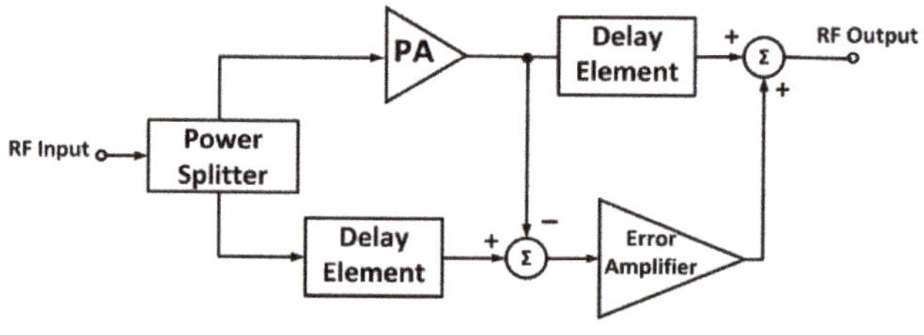

Figure 22. Block diagram of the feed-forward-based PA.

The feed-forward approach was realized in a 24 GHz cascode CMOS PA, as demonstrated by Y. H. Chen et al. (2015) [121]. The PA employed a successive second-order intermodulation (IMD2) feed-forward cancellation. The main path comprised a cascode PA operated in class AB, while the auxiliary path comprised successive IMD2-generating common-source amplifiers. The transistors in the common-source amplifiers were operated in the class B region, in which it detected the fundamental signals and generated the maximum IMD2 components and moderate fundamental components. Further, the transistor in the final stage generated IMD3 components that were utilized to cancel the main path's IMD3 components. A diode-connector-based level shifter was employed at the final stage in order to realize the biasing voltage in class B and also to allow the very-low-frequency signals to pass through. The PA was capable of effectively mitigating the IMD3 by reducing the third-order transconductance, which is the major contributor to IMD3. The IMD3 value was enhanced by 20 dB at the "sweet spot" located at an output power of 5 dBm, with the auxiliary stage being active. The IMD3 measurement was carried out with a 1 MHz spacing two-tone test at 24 GHz, while the ACPR measurement was completed with 64 QAM, at a 1 MHz symbol rate.

4.3. Linear Amplification Using Non-Linear Components Technique

The linear amplification using non-linear component (LINC) is usually employed with highly efficient PAs such as class C, D, E, or F. In LINC, the modulated input signal is split into constant envelope and phase-modulated signals. These signals are independently amplified by two PAs with the same features. The two amplified signals are later combined together at the output with fewer distortions [122]. The separated signals are constant envelope, thus they are insensitive to the non-linearity of the PAs. Therefore, the employment of non-linear switching-mode PAs is advantageous when implemented in LINC techniques. Figure 23 illustrates the typical block diagram of a LINC technique.

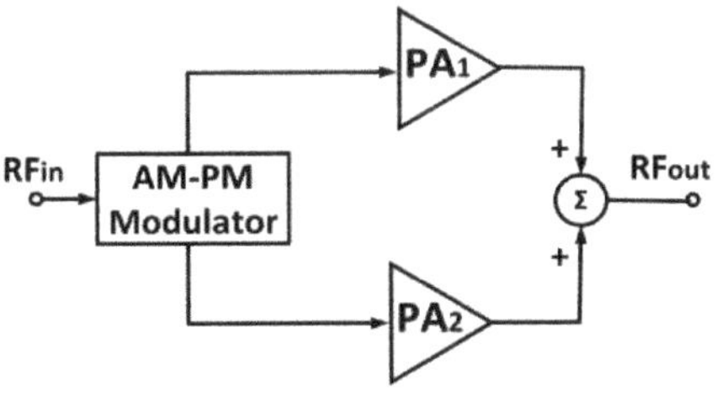

Figure 23. Block diagram of LINC technique.

H. Lee et al. (2010) presented a multi-mode LINC PA that was employed with a differential two-pair class AB CMOS PA [123]. The PA comprised two pairs of driver stages and main power stages that employed a cascode structure and stabilizing feedback circuits. The multi-mode LINC PA was capable of enhancing its linear output power by 7% and its efficiency by 8% compared to the conventional method. H. Lee et al. (2013) also presented a hybrid polar-LINC PA that employed a combination of polar and LINC mechanisms [124]. This hybrid polar-LINC PA was configured in a differential structure comprising two pairs of driver stages and power stages and an output combiner. The driver and power stages were both based in class E operation. The proposed power combiner comprised a transmission line transformer (TLT) with isolation resistors. The employed isolated TLT avoided distortion induced by the mutual load-pulling effects, which occur in non-isolated combiners [125]. The isolated combiner allowed a more linear and precise signal combination that resulted in ACLRs of −36 dBc and −47 dBc under 5 MHz and 10 MHz WCDMA signals, respectively.

4.4. *Pre-Distortion Technique*

Pre-distortion employs an approach of intentionally introducing distortions in terms of amplitude and phase that are in contrast to a PA's distortion [126]. This results in the cancellation of distortions and, thus, enhances the linearity. The pre-distortion technique provides a wide-bandwidth operation due to its open-loop feature, which makes it advantageous over feedback techniques [127]. Pre-distortion is also advantageous over the feed-forward technique due to its simplicity and low power consumption [128]. The pre-distorter mechanism is implemented at the input of the PA, as shown in Figure 24. Pre-distortion is highly flexible, effortlessly reconfigurable, and suitable for wideband application. However, the PA's characteristics' variation is subjected to component tolerances. The pre-distortion mechanisms can be employed in an analog or digital manner.

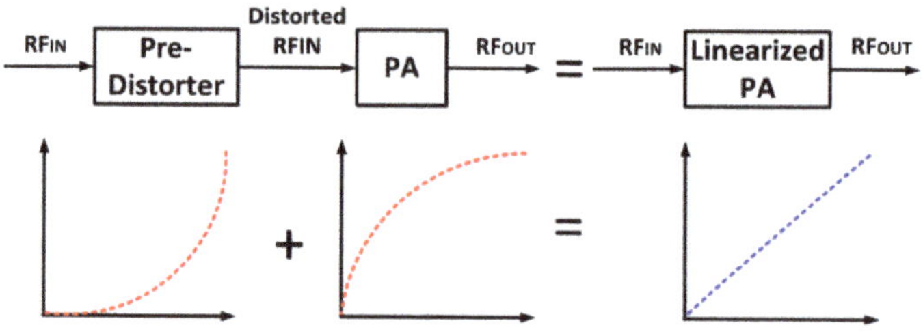

Figure 24. The concept of pre-distortion.

4.4.1. Analog Pre-Distortion Technique

Analog pre-distortion (APD) employs a non-linear mode device to pre-distort the input signal prior the PA. RF pre-distortion is usually implemented in an analog mechanism due to the complexity of the sampling frequency needed to be digitally processed.

K. Y. Son et al. (2012) proposed a CMOS PA with an integrated digitally controlled APD [129]. This APD had two external digital control words, which are the I-path and Q-path control words. These digital control words were utilized to control the AM–AM and AM–PM distortions in the PA. The driver stage of the PA was employed in a vector modulator manner, so the gain and phase can be varied via the vector sum method [130,131]. The linear output power of the PA was enhanced by 6.5% with this approach. Furthermore, a 15 GHz DPA with an integrated APD as presented by N. Rostomyan et al. (2018) [132]. The main and peaking paths were both constructed with two-stage amplifiers. In the input and output power combiners, high-pass π-networks were utilized due to their advantages in terms of stability and a compact area compared to a $\lambda/4$ transmission line. The APD in this design focused on the gain non-linearity issue of a symmetrical DPA. The linear output power and gain flatness were achieved by controlling the supply and bias voltages of the envelope detector employed in the PA. An improvement of 9.1% in the linear output power was observed via the implementation of the proposed APD.

Furthermore, S. Mariappan et al. (2021) presented a CMOS PA with a wideband pre-distortion (WPD) mechanism that consisted of a hybrid feedback with both active and passive networks [133]. The WPD provided both gain and phase cancellation schemes that aid in mitigating the third-order intermodulation product. The WPD exhibited a capacitive response that is in opposition to the main PA's inductive response. An active load was implemented at the WPD instead of an RF choke, which generated the capacitive response. The proposed CMOS PA with WPD achieved a maximum output power of 24 dBm, with a peak PAE of 35.5% at 2.45 GHz. The achieved linear output power was 20 dBm across the operating frequency bandwidth of 0.8–3.3 GHz. S. Mariappan et al. (2021) also presented a CMOS PA with a broadband pre-distortion (BPD) that used the same aforementioned mechanism [134]. However, this CMOS PA as implemented on a Roger's RO4000/FR4 PCB, which achieved a maximum linear output power of 21 dBm across the frequency bandwidth of 0.4–2.8 GHz. The peak efficiency achieved is more than 35%.

S. Mariappan et al. (2021) proposed another CMOS PA with an integrated digitally assisted analog pre-distorter (DAAPD), which offers a reconfigurable linearization scheme [135]. The DAAPD was utilized to tune the interstage load impedance between the driver stage and the main stage. The load impedance was reconfigured by varying the transconductance of the driver amplifier and its respective active load. The DAAPD consisted of a digital bias controller that comprised a seven-stage voltage generator with a band-gap reference voltage, buffer, and op-amp, which was employed to provide the biasing voltage that controls the transconductance. The DAAPD was also employed to optimize the process–voltage–temperature variation in the CMOS PA that contributes to its robustness. The DAAPD generated a capacitive response as opposed to the inductive response of the main PA that enhances the linearity. This CMOS PA exhibited a maximum output power of 27–28 dBm across the frequency bandwidth of 1.7–2.7 GHz. The linear output power achieved was 24–25 dBm.

Based on the aforementioned design, S. Mariappan et al. (2023) further improved the presented design by proposing a reconfigurable CMOS PA with digital linearizers and a tunable-output impedance matching network for PVT robustness [136]. The digital linearizer was implemented at the driver stage of the CMOS PA, which acted as the wideband pre-distorter. The tunable-output matching network was realized via an interleaved planar transformer with a switched capacitor at both the primary and secondary windings. The switched capacitor was capable of maintaining the efficiency and the output power of the CMOS PA across its operating frequency bandwidth as well as PVT, while the digital linearizer conducted the linearization tuning. The proposed mechanisms for the CMOS PA achieved a linear output power of 24–25 dBm and a PAE of 34.5%–38.8% across the frequency of 1.7–2.7 GHz.

In addition, J. Tsai (2023) proposed a CMOS PA with a high-power APD constructed with a cascode cold-FET [137]. The CMOS PA was comprised of three differential main PA units, a driver amplifier, a two-stage dual-radial power splitter/combiner and a pre-distorter as the linearizer. The cascode cold-FET had a higher breakdown voltage that can sustain a higher input power and generated a larger gain expansion region compared to conventional common-source cold-FET implementation. This allowed the APD to conduct linearization at a higher-output power region and further enhanced the linear output power of the proposed watt-level CMOS PA. The proposed CMOS PA operated at 5.3 GHz, with a maximum output power of 30.1 dBm and a peak PAE of 18%. The achieved linear output power was 21 dBm after linearization was conducted.

4.4.2. Digital Pre-Distortion Technique

The digital pre-distortion (DPD) technique is gaining popularity due to the rapid development of digital signal processing (DSP) [138]. Due to the limitations in terms of complexity, cost, and efficiency, the DSP-based DPD is restricted to less than a 100 MHz operation bandwidth. The DPD mechanism is commonly implemented with direct and indirect learning architectures [139–141].

Y. Cho et al. (2014) presented a CMOS DPD PA with an off-chip open-loop DPD algorithm that only uses a single look-up-table (LUT) [142]. The DPD was employed for a large dynamic range of output power via a LUT-based average output power control. An enhancement of 8 dB was achieved in ACPR by utilizing the proposed DPD.

Moreover, a two-core digital CMOS PA with on-chip integrated DPD was presented by A. Wong et al. (2017) [143]. A polar architecture and an on-chip 1-D LUT were employed in the proposed design. The digital PA adopted a double cascode configuration. The NAND gates were employed to drive the device's transconductance with inputs of amplitude and phase switching at the rate of the carrier frequency. The amplitude data were obtained via the on-chip DSP in four 12-bit data streams at a rate of one-fourth of the carrier frequency. On the other hand, the phase data were obtained via two sets of four parallel 10-bit data streams that were converted into two series of 10-bit data streams before they were supplied into the I-Q DAC. A piecewise linear DPD was employed in this design, which achieved linear output power of 26.8 dBm and 25 dBm for the 20 MHz 802.11g and 40 MHz 802.11n signals, respectively.

4.5. Adaptive Biasing Technique

Linearization can also be conducted by controlling the transconductance of the CMOS PA. Since this tranconductance is directly dependent on the gate voltage of the transistor, adaptive biassing circuits are extremely beneficial for this implementation. They only provide a variable bias current when an input signal is present. In the absence of such input, they provide a constant, precisely controllable quiescent current. In this manner, the quiescent power consumption is decreased without compromising the transient performance.

W. Kim et al. (2014) proposed a CMOS PA with an adaptive biasing technique for EDGE/GSM application [144]. To improve the AM–AM characteristics, this CMOS PA was adaptively biased according to its input power level. The adaptive bias circuit was designed to enhance bias voltages when the input signals were large, thereby increasing the 1 dB output power in proximity to the peak power. Also, the non-linear gate–drain capacitance (C_{gd}) of the power transistors, which is a primary source of AM–PM non-linearity, was effectively linearized by adding a series capacitor. The proposed CMOS PA achieved PAEs of 22% and 23% at linear output powers of 28.5 dBm and 27.5 dBm, at 870MHz and 1.8 GHz, respectively.

H. Kim et al. (2015) introduced a CMOS PA with the implementation of a dynamic bias switching method via digital control [145]. The digital signal processing unit generated the control signal for the bias switching operation based on the amplitude of the envelope signal. The dynamic bias switching circuit provided dual supply voltages to the PA's drain. The dynamic bias switching circuit produced a low supply voltage through the use of a DC–DC converter, whereas the high voltage can be directly sourced from the power supply. The compensation of the gain difference between low- and high-bias voltage conditions was implemented to enhance the linearity of the CMOS PA. The optimization of the threshold voltage for the envelope signal and the low supply voltage level was carried out through analytical means to achieve the highest possible efficiency without trading off its linearity. This was achieved by utilizing the envelope statistics of the LTE signal. This CMOS PA achieved a linear output power of 22 dBm, with a peak PAE of 34.5% at 1.75 GHz.

B. Kim et al. (2016) proposed a CMOS PA using the reconfigurable adaptive power cell technique. This CMOS PA was configured in a cascode structure that consisted of four differently biased power cells to achieve high linearity [146]. Out of the four power cells, two of them were shut off to operate the CMOS PA at a high frequency. This made the CMOS PA operate at a dual-frequency band. The power cells were designed with different channel widths that give different input average capacitances. Therefore, the total average capacitance variation with regard to the input power may become relatively insignificant as a result of mutual cancelling, which contributes to enhancing its linearity. The maximum output power attained was 27.8 dBm, with a peak PAE of 52.5%. This CMOS PA achieved a linear output power of 17.5 dBm at 650 MHz.

Furthermore, J. Ren et al. (2020) introduced a CMOS PA with an adaptive bias circuit that generated a precise biasing voltage according to the input envelope RF signal [147]. This CMOS PA also employed harmonic termination circuits at the gate of the common-gate PA and the center-tap node of the output transformer. The harmonic termination aided in shorting the second-harmonic components to the ground, which further enhanced the linearity. This CMOS PA achieved a maximum output power of 30.3 dBm, with a peak PAE of 43.1% at 2.4 GHz. The achieved linear output power was 23.5 dBm, with 18.1% linear efficiency.

S. Mariappan et al. (2021) presented a dual-stage CMOS PA with adaptive biasing in which the gate voltage of the driver amplifier was adaptively tuned to be in opposition to the third-order transconductance response compared to the main PA [148]. The biasing voltage was supplied through a current-mirror configured biasing circuit. The opposite third-order transconductances between the stages cancel each other by assisting in mitigating the third-order intermodulation product. This proposed CMOS PA achieved a linear output power of 19 dBm, with a respective PAE of 29% at the 2.45 GHz operating frequency.

A.S. Rawat et al. (2022) presented a CMOS PA with adaptive bias technique for LORA application operating at 919–923 MHz [149]. The driver stage of the CMOS PA utilized a bias voltage modulation mechanism, while the main PA stage used a split bias mechanism to achieve different power level modes. A bias circuit controlled the channel current of the driver and the main PA via the applied gate voltage. The main PA was split into two units, in which each of them was adaptively biased to achieve the optimized linearity performance. A linear output power of 21 dBm, with a respective PAE of 29% at 920 MHz, was obtained. A. S. Rawat et al. (2023) also presented another CMOS PA with a switching mechanism for adaptive-device-sizing biasing [150]. This CMOS PA also employed a high Q-factor compact inductor for efficiency enhancement at linear output power. The adaptive device sizing was implemented to enable the low-power and high-power modes for the PA. The switching of the modes was selected through a NMOS pass transistor. Two individual bias circuits were utilized to conduct the adaptive biasing for the main PA stage. For the low-power mode, only one of the bias circuits was turned ON. In contrast, both bias circuits were turned ON for the high-power mode operation. Thus, the linear output power can be increased with minimum trade-off with efficiency via this method. The proposed CMOS PA achieved a linear output power of 21.5 dBm, with a respective PAE of 22%. Table 3 summarizes the recent state-of-the-art CMOS PAs with linearity enhancement techniques.

Table 3. Summary of the recent state-of-the-art CMOS PAs with linearity enhancement techniques.

Architecture	Year	Tech (nm)	V_{DD} (V)	Freq. (GHz)	Linear. P_{out} (dBm)	Linear PAE (%)	Ref.
Feedback/Feedforward	2016	130	1.2	0.868	10.4	29.0	[116] +
	2013	150	3.3	1.75	24.0	16.0	[118]
	2013	65	2.0	2.0	20.4	32.3	[119]
	2015	180	3.6	24.0	14.6	12.6	[121]
LINC	2010	180	3.3	1.95	26.0	20.0	[123]
	2013	130	3.5	1.95	28.5	29.6	[124]
Pre-Distortion	2012	180	3.5	1.95	29.1	41.1	[129]
	2018	45	4.8	15.0	20.6	21.8	[132]
	2021	180	3.3	0.8–3.3	20.0	34.0	[133]
	2021	180	3.3	0.4–2.8	21.0	34.6	[134]
	2021	180	3.3	1.7–2.7	25.0	38.0	[135]
	2023	180	3.3	1.7–2.7	25.1	38.8	[136]
	2023	180	3.3	5.3	21.0	6.0	[137]
	2014	180	3.5	1.7–2.0	28.0	42.0	[142]
	2017	28	3.3	2.4	26.8	21.2	[143]
Adaptive Biasing	2014	180	3.5	0.87	28.5	22.0	[144]
	2015	180	3.3	1.75	22.0	39.3	[145]
	2016	40	-	0.65	17.5	10.5	[146]
	2020	130	2.5	2.4	23.5	18.1	[147]
	2021	180	3.3	2.45	19.0	29.0	[148]
	2022	180	3.3	0.92	21.0	29.0	[149]
	2023	130	3.3	0.92	21.5	22.0	[150]

+ simulated.

Figure 25 depicts the plot of the linear PAE across the achieved linear output power by the aforementioned linearization techniques. It can be observed that most of the employed techniques achieved a linear output power of more than 20 dBm. Most of the CMOS PAs that employ the analog predistortion technique were capable of achieving more than 30% linear PAE. Also, it can be deduced that the CMOS PAs were still limited in delivering a watt-level linear output power, since the PAs were required to operate at a backed-off output power to sustain the linear transmission.

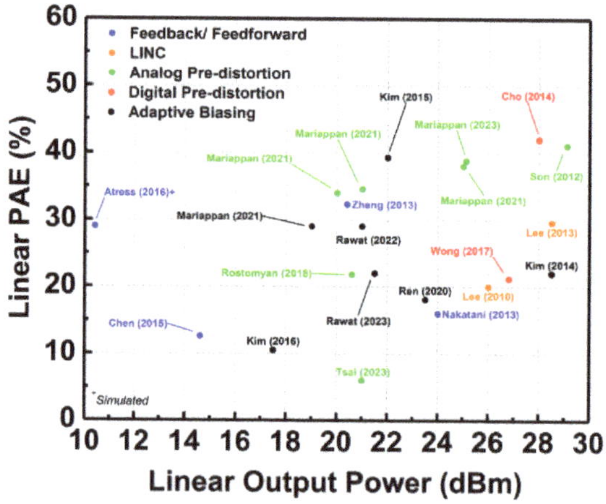

Figure 25. Linear PAE across linear output power for the proposed linearity enhancement techniques [116,118,119,121,123,124,129,132–137,142–150].

5. Bandwidth Enhancement Techniques

A transistor's gain relatively declines with increasing frequency, with an estimated factor or roll-off slope in the order of 4–6 dB per octave. Thus, in a wideband PA, it is vital to extend the gain response across the required bandwidth with minimum gain deviation. Moreover, the PA's matching network needs to be designed with a low insertion loss slope for the gain compensation in order to achieve a flat gain as well as high efficiency and linearity. Output matching network is vital in a PA design for realizing optimum output power, efficiency, and linearity. The input and interstage matching networks provide the intended bandwidth extension. Commonly, the input or driver stage of a wideband PA is employed with a negative feedback circuit to minimize the gain loss at higher frequencies [151]. Several bandwidth enhancement techniques were introduced, which include distributed, cascode, feedback, and switched-capacitor PAs.

5.1. Distributed Technique

The distributed amplifier (DA) achieves a high bandwidth by subsuming the transistor's gate-source capacitance (C_{gs}) into an artificial transmission line at the input of the PA [152]. The gate and drain line inductors create artificial lines with the C_{gs} and C_{ds} of the amplifiers. Usually, the inductors are replaced with high-impedance microstrip lines in order to achieve low noise and high output power from the PAs. Figure 26 depicts a simplified n-section DA [153].

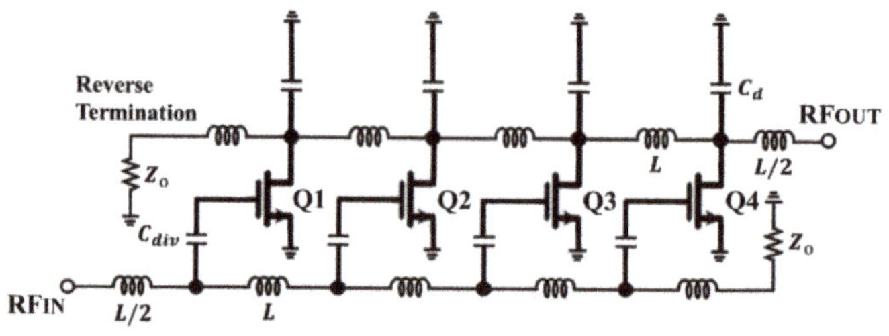

Figure 26. Simplified n-section DA.

The signal is amplified via its transconductance and accumulates along the drain line as the signal passes through the PA. The DA method produces a wide bandwidth as a result of the low-pass structures, which produce high cut-off frequencies. The impedance of the input line is defined as

$$Z_{og} = \left(\frac{L_g}{C_{ds}}\right)^2 \quad (19)$$

In contrast, the impedance of the drain line is defined as

$$Z_{od} = \left(\frac{L_d}{C_{ds}}\right)^{\frac{1}{2}} \quad (20)$$

Since the gate capacitance restricts the achievable bandwidth of the DA, the cut-off frequency is defined as

$$f_c = \frac{1}{\pi Z_{og} C_{gs}} \quad (21)$$

Figure 27 shows the frequency response comparison of a DA and PA with lumped elements.

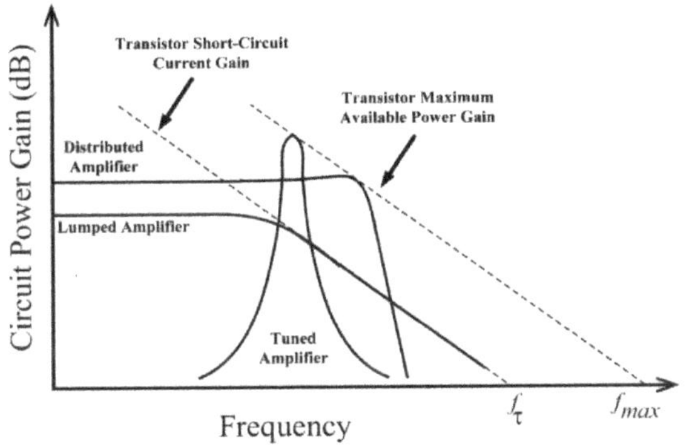

Figure 27. Frequency response comparison.

C. Hsiao et al. (2013) proposed a CMOS DA that employs a gate–drain transformer feedback method [154]. This employed feedback method permits the reuse of the traveling signal in order to attain a large gain bandwidth while preserving the low power consumption of the DA. The utilized transformer is miniaturized via a folded transmission lines configuration. A patterned ground shield (PGS) is employed to enhance the quality factor of the transformer. The transformer also has a well-controlled feedback coupling coefficient. This proposed DA achieved a 3 dB bandwidth of 61.3 GHz, with power consumption under 60 mW.

Furthermore, P. Chen et al. (2014) presented a DC-80 GHz compact CMOS DA based on a conventional DA with the gain cell of a cascaded single-stage DA (CSSDA) [155]. To make it area efficient, the artificial transmission lines (ATL) were employed with a micro-strip line instead of a coplanar waveguide. The inter-digital capacitors utilized in this design were optimized for wideband frequency response. The achieved 3 dB bandwidth was 80 GHz, with power consumption under 90 mW.

Y. Zhang et al. (2017) proposed a CMOS DA comprised of two three-stage DAs integrated with an input ATL that served as a preamplifier for broadband impedance matching as well as gain enhancement [156]. In addition, two four-stage DAs were integrated with the output ATLs that served as a medium PA to exhibit the high output power for a wide bandwidth. This proposed DA had a bandwidth operation from 2 to 22 GHz.

M. M. Tarar et al. (2017) presented a wideband stacked distributed CMOS PA [157]. A four-transistor stack was employed to achieve a high output swing and gain. Voltage alignment was achieved by allowing a small AC swing at the gate to be derived from the voltage division between C_{gs} and the external gate capacitance. The four-transistor stack was replicated in the distributed configuration, which contributed to the wideband operation. This PA was capable of operating with a bandwidth from 2 to 16 GHz.

5.2. Cascode Technique

A cascode configuration is employed by connecting common-source (CS) and common-gate (CG) transistors in series, as illustrated in Figure 28. The cascode topology is widely utilized in PA designs to enhance the gain and bandwidth [158]. The cascode structure has several advantages, which include high output impedance, high reverse isolation, and less sensitivity to the output resistance of the transistors. High reverse isolation is contributed to by the low Miller feedback capacitance. Moreover, the reduction in Miller capacitance exhibits a 3–4 dB higher gain and increased bandwidth compared to CS structures [159].

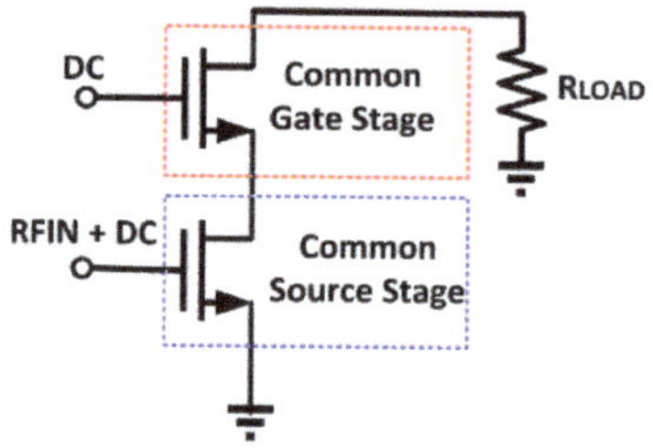

Figure 28. Cascode PA configuration.

In a CS configured device, the reduction in the Miller capacitance in the gate–drain parasitic capacitance (C_{gd}) generates a frequency pole that makes a small signal gain roll down. The Miller capacitance in the C_{gd} for a CS amplifier with a voltage gain of $A_v = -g_m$ R_{load} is given by

$$C_{M=}C_{gd}(1 + A_v) \tag{22}$$

The increase in Miller capacitance reduces the bandwidth of the PA. When the cascode transistor is integrated as a load to the drain of the CS amplifier, the Miller effect is eliminated regardless of the load impedance. This feature enhances the bandwidth of the PA. In a two-identical-cascodes device, the small signal open loop voltage gain is defined as [160]

$$A_{vo} = \frac{v_o}{v_i} = -g_{m1}r_{o1}x(1 + g_{m2}r_{ro2}) \approx -g_{m1}r_{o1}g_{m2}r_{o2} \approx -g_m^2 r_o^2 \tag{23}$$

where g_m is the small signal transconductance gain, and r_o is the transistor output impedance. S. Leuschner et al. (2011) proposed a multi-band CMOS PA with a cascode configuration output stage for WCDMA application [161]. A two-stage interstage matching network was utilized to achieve a wideband operation of more than 300 MHz. The cascode PA achieved an efficiency of more than 45% across the operating bandwidth. H. Jeon et al. (2013) proposed a CMOS multi-stage cascode PA with a feedback bias technique [162]. The feedback bias technique was utilized to enhance the linearity and reliability across a wideband frequency range. By utilizing the large parasitic capacitance and low substrate resistivity of the CMOS process, the signal swings were coupled between the ports of the transistors. The leakage signals at the gate of CG structure were employed for the negative feedback. The proposed cascode PA achieved a 3 dB bandwidth of 1.5–2.1 GHz.

H. Wu et al. (2016) proposed a two-stage three-stacked PA with a wideband gain [163]. Wideband load impedance matching was achieved by utilizing modified stacked field-effect transistors with a resistive feedback. To realize the wideband gain frequency response, the compensation between the driver amplifier's expanding gain response and the last-stage amplifier's compressing gain response was utilized. The PA achieved a bandwidth of 6.4 GHz from 0.1 to 6.5 GHz. It delivered a maximum output power of 22–24.3 dBm, with a peak efficiency of 13–20%. The achieved efficiency was quite low due to the three-stacked FETs employed at the driver and the last stages, which consume 39.1% and 41% of power, respectively.

In addition, S. Kang et al. (2018) presented a cascode CMOS PA with integrated dynamic body linearizers [164]. The dynamic body linearizers were employed based on the envelope signal injection to the bulk of the CS and CG configured transistors. The dynamic body linearizers mitigated the parasitic capacitances, and these capacitances could limit the modulation bandwidth due to the delays cause by them between the dynamic bias linearizer's output and the input envelope signals. This cascode PA was capable of performing from 1.7 to 2.0 GHz.

T. Guo et al. (2022) proposed a cascode CMOS PA with a split push–pull configuration at both the driver and main PA stages [165]. This CMOS PA also employed floating-body transistors in both stages. An on-chip bandgap voltage reference was implemented in the design in which it generated the temperature-independent biasing voltage for both the driver and main PA. The design also utilized on-chip transformer baluns for power splitting and combining. The baluns also acted as the wideband impedance matching for the input and output of the CMOS PA. This proposed CMOS PA achieved an operating frequency bandwidth of 4.3–6.4 GHz. This PA achieved a maximum output of 22.2 dBm at 4.9 GHz and a peak PAE of 28% at 4.7 GHz.

5.3. Shunt Feedback Technique

Resistive shunt feedback is a common technique employed to provide wideband matching in PAs. A PA with a resistive feedback provides a good output reflection coefficient. The common practice is to implement a negative resistor feedback from the drain to the gate, as depicted in Figure 29 [166].

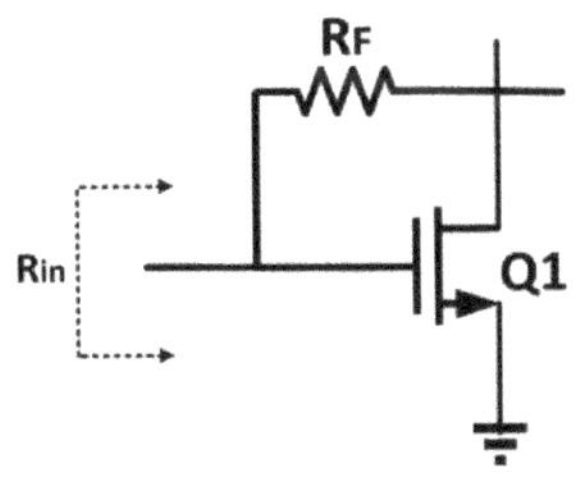

Figure 29. Configuration of a resistive feedback.

Referring to Figure 29, for a PA with a gain of A_v, the input resistance, R_{in}, is given by

$$R_{in} = \frac{R_F}{1 + A_v} \tag{24}$$

A small R_F provides excellent matching but with the cost of a reduced gain due to the signal feedback through this path. Contrarily, a large R_F provides a high gain but reduces the effect of the feedback.

A straightforward feedback resistor is not adequate when the bandwidth requirements are very large. Thus, some combination of passive or active elements is required in the feedback path. Considering a feedback employed in a CS configured amplifier, a small signal model is shown in Figure 30. Based on Figure 30, the input impedance of a CS configured amplifier is given as

$$Z'_{in} = sL_{bond} + \frac{1}{sC_{gs}} + \frac{g_m L_{bond}}{C_{gs} + C_{gd}} \tag{25}$$

$$g_{m,eff} = \frac{g_m v_{gs}}{v_{in}} = \frac{g_m}{1 + sg_m L_{bond}} \tag{26}$$

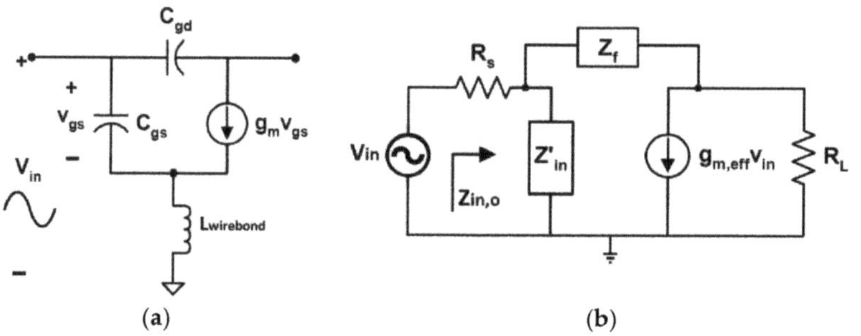

Figure 30. (**a**) Small signal model of CS; (**b**) small signal model of CS with feedback.

After employing a feedback into the CS amplifier, the input and output impedances and their corresponding gains are defined as

$$Z_{in,o} = \frac{Z'_{in} Z_f}{Z_f + (A_v - 1)Z'_{in}} \tag{27}$$

$$Z_{out,o} = \frac{Z_f (R_s \parallel Z'_{in})}{1 + g_{m,eff}(R_s \parallel Z'_{in})} \tag{28}$$

$$A_v = \frac{R_L + g_{m,eff} Z_f R_L}{R_L - Z_f} \tag{29}$$

Since Z'_{in} and $g_{m,eff}$ are frequency-dependent parameters, the characteristics of Z_f also correspondingly vary across the frequency band.

S. A. Z. Murad et al. (2010) presented a CMOS PA with a resistive feedback to achieve a wideband flat gain [167]. This PA also adopted a cascode topology with a current reuse method that was utilized to increase the gain at the upper end of the operating band. An inter-stage inductor was employed in between the first and second stages of the PA, which further contributed to the gain flatness. This proposed PA achieved a flat gain with ±0.8 dB deviation from 3.1 to 4.8 GHz and a 3 dB bandwidth of 2.8–5.0 GHz.

S. A. Z. Murad et al. (2010) also presented a wideband CMOS PA, which achieved a flat gain by employing a shunt–shunt feedback [168]. This PA utilized a current reuse technique in the first stage to enhance the gain at the upper band of the operating frequency. The second stage of the PA consisted of a shunt and series peaking inductors with a resistive feedback and a shunt–shunt feedback. The resonance circuit in this PA provided a narrow band characteristic that increased the gain at the upper end of the operating frequency (7 GHz). A shunt inductor was inductively peaking at the low frequency region (3 GHz), thus realizing a wide-bandwidth gain. A series peaking inductor was employed between the first and second stages in order to increase the middle band's gain. This contributed to the gain flatness across the wide bandwidth. The employed resistive feedback aided in further flattening the gain. A shunt–shunt feedback using a resistor and a capacitor further increased the bandwidth and improved the wideband output matching. This PA achieved a flat gain with ±0.5 dB deviation from 3.0 to 7.0 GHz as well as a 3 dB bandwidth of 2.8–7.4 GHz.

In addition, N. Ginzberg et al. (2023) presented a wideband CMOS PA that implemented a cascode topology with a resistive shunt–shunt feedback [169]. This CMOS PA was biased in the class B operation region. The design adopted simultaneous input and output matching via the feedback resistor. The biasing network was also resistively implemented in order to achieve a wideband response. The shunt–shunt resistive feedback technique was required to conduct impedance matching as well as stability. In order to achieve both unconditional stability and optimum input matching, a 15 Ω resistor was employed at the input. This proposed CMOS PA achieved an operating frequency bandwidth of 0.33–2.5 GHz. The maximum output power delivered was 21.5 dBm, with a peak PAE of 52.4% at 1 GHz.

5.4. Switched Capacitor Technique

A switched capacitor PA (SCPA) has been extensively trending in recent years, and it employs an array of capacitors that are tuned between the supply and ground [170–177]. Figure 31 shows the schematic of the basic switched capacitor (SC) architecture [178].

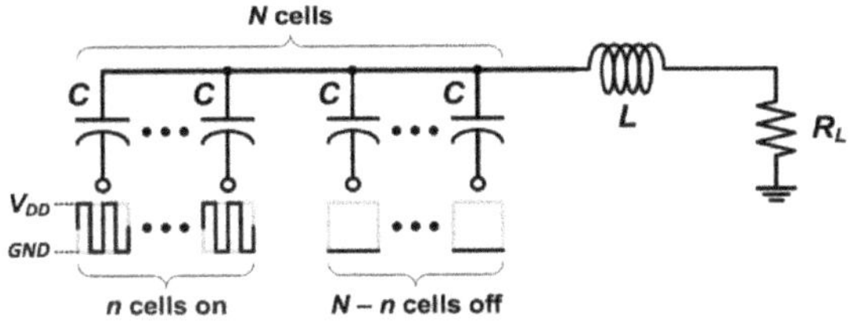

Figure 31. Schematic of the basic switched capacitor (SC) architecture.

The amplitude of the signal is created by a capacitive voltage division between the switched-on capacitors and switched-off capacitors as follows [179]:

$$Vout = \frac{C_{ON}}{C_{ON} + C_{OFF}} \cdot V_{DD} = \frac{C_{ON}}{C} \cdot V_{DD} \tag{30}$$

where C_{ON} is the equivalent capacitance of the switched-on capacitors, C_{OFF} is the equivalent capacitance of the switched-off capacitors, and C is the total array capacitance. The capacitor array is integrated with a band-pass matching network that resonates the total capacitance at the frequency of application.

The output power is obtained by considering the output voltage as a fundamental component of the Fourier-series expansion of a square wave signal. The output voltage and power are given as

$$Vout = \left(\frac{2}{\pi}\right)\left(\frac{n}{N}\right)V_{DD} \tag{31}$$

$$P_{out} = \left(\frac{2}{\pi^2}\right)\left(\frac{n}{N}\right)^2 \frac{V_{DD}^2}{R_{opt}} \tag{32}$$

where n is the number of switched-on capacitors, and N is the total capacitance. The PAE of the SCPA is given by

$$PAE = \frac{P_{out}}{P_{DC,SCPA} + P_{SW,in}} \tag{33}$$

where $P_{SW,in}$ is the dynamic losses of the SCPA core drivers and the clock distribution network, and $P_{DC,SCPA}$ is the average power consumption of the PA core. $P_{DC,SCPA}$ also takes into account the ohmic losses of the switch resistances as well as the dynamic power required for the charging and discharging of the capacitor array [180].

S. Goswami et al. (2014) presented a CMOS PA with a frequency agile solution that was tunable over a wide-bandwidth frequency [181]. This CMOS PA utilized a class D configured out-phasing technique with an integrated tunable matching network. The tunable matching network was realized via a series combining a transformer and a digitally tuned capacitor bank. The unit capacitance was employed in unary weighted mode. The proposed PA was implemented using 45 nm CMOS SOI technology with a supply headroom of 1.8 V. It was capable of operating from 1.44 to 3.41 GHz, with a bandwidth of 1.97 GHz. This PA delivered a maximum output power of 27–28.2 dBm across the frequency, with a peak efficiency of 25–30%. The achieved linear output power was 21.6–23.4 dBm, with an efficiency of 9.8–17.4% across the frequency of operation.

S. Yoo et al. (2011) presented an SCPA for EER or polar transmitters without the need for a supply modulator [182]. A baseband DSP was employed to generate a polar-modulated signal, and the number of capacitors to be switched were controlled by the digital envelope input signal to a thermometer decoder. This SCPA can be considered as a capacitive power combination of multiple switching PAs in which the output power is the combination of the charge redistribution on the capacitors. This proposed PA was capable of operating from 1.8 to 2.8 GHz.

Furthermore, W. Yuan et al. (2016) proposed a digital-based class G quadrature SCPA, which combined the I and Q signals on a shared capacitor array [183]. This proposed Q-SCPA did not need a wideband phase modulator or a delay matching circuitry, unlike its polar or EER counterparts. The capacitor array was subdivided in such a way that half of the array was located in the path of the individual 1/Q paths. The subarrays were also divided into unit capacitors. The unit capacitors had individual driver chains that could be switched on and off. A 3 dB bandwidth of 400 MHz from 1.9 to 2.3 GHz was achieved by the proposed SCPA.

Z. Bai et al. (2018) proposed a multiphase SCPA consisting of C-2C split array capacitors [184]. This SCPA was designed for 16-bit resolution for additional linearization or calibration states using DPD. C-2C arrays are extensively utilized due to the low ratio of the maximum capacitance to the minimum capacitance, which offers a better impedance matching [185,186]. However, this comes with the expense of increased sensitivity to parasitic effects [187,188]. The C-2C method consumes less area compared to binary weight arrays because C-2C arrays linearly increase with the resolution. The proposed SCPA achieved a 3 dB bandwidth of 700 MHz from a center frequency of 1.8 GHz. Moreover, Z. Bai et al. (2019) also presented a frequency-tunable SCPA that achieved an operating bandwidth of more than 1 GHz [189]. This SCPA was employed in series with an inductor and a digitally programmable capacitor (DPC). The frequency tuning of the SCPA resonance was conducted by the DPC. The SCPA comprised an 8-bit array that consisted of 4-unary and 4-binary weighted bits. The switch and driver stages were employed as custom cells that can be tiled with nearby cells in order to mitigate parasitic capacitance. This proposed SCPA was capable of tuning its frequency from 1.4 to 2.5 GHz of bandwidth, while maintaining an output power of 20–21.7 dBm.

I. Raja et al. (2020) proposed a reconfigurable RF transmitter with an integrated tunable class E PA [190]. A DSP was utilized to minimize the sampling spurs up to the harmonics of the carrier, which avoided the requirement for sharp RF filters. Other than that, a DPD was also employed as a linearizer for the class E PA. Clock correction circuitry was implemented for rectifying the duty cycle and quadrature errors. In the impedance matching network, the capacitors were made tunable to provide wave shaping across the frequency range, while using a single, fixed-series inductor. The switching transistors were the control bits for the tunable capacitor bank. The capacitors were segmented into smaller capacitors in order to achieve an effective tuning. This proposed PA had a bandwidth of 1.75 GHz, from 0.75 to 2.5 GHz. This PA delivered a maximum linear output power of 6.5–13.0 dBm, with a linear efficiency of 5–27%, when tested with a 25-kS/s 16-QAM signal. In addition, the finite impulse response (FIR) filters, DPD segments, and the RF choke inductor were employed off-chip in this design. Table 4 summarizes the recent state-of-the-art CMOS PAs with bandwidth enhancement techniques.

Figure 32 shows the summary of the frequency bandwidths achieved by the proposed bandwidth enhancement techniques. Based on Figure 32, the largest bandwidth is achieved by the distributed technique, which has a range from the DC level to 80 GHz. However, the other performance parameters of the PA such as its output power and PAE are very low for this technique. The other techniques are capable of achieving an operating bandwidth of more than 1 GHz, with an optimum output power and PAE for its given application. The switched-capacitor-based CMOS PA has the most flexibility and configurability in tuning its parameters according to its need, since its impedance can be varied.

Table 4. Summary of the recent state-of-the-art CMOS PAs with bandwidth enhancement techniques.

Architecture	Year	Tech (nm)	V_{DD} (V)	Freq. (GHz)	Bandwidth (GHz)	Ref.
Distributed	2013	90	2.2	DC–61.3	61.3	[154]
	2014	40	1.7	DC–80.0	80.0	[155]
	2017	180	2.8	1.0–23.8	22.8	[156]
	2017	130	3.5	2.0–16.0	14.0	[157]
Cascode	2011	65	3.4	1.64–1.95	0.31	[161]
	2013	180	3.4	1.5–2.1	0.6	[162]
	2016	180	6.0	0.1–6.5	6.4	[163]
	2018	180	3.3	1.7–2.0	0.3	[164]
	2022	130	3.3	4.3–6.4	2.1	[165]
Shunt Feedback	2010	180	1.8	2.6–5.0	2.4	[167]
	2010	180	1.8	3.0–7.0	4.0	[168]
	2023	65	3.3	0.33–2.5	2.17	[169]
Adaptive Biasing	2014	45	2.0	1.3–3.3	2.0	[181]
	2011	90	2.5	1.8–2.8	1.0	[182]
	2016	65	2.4	1.8–2.3	0.5	[183]
	2018	65	2.4	1.4–2.1	0.7	[184]
	2019	65	2.4	1.4–2.5	0.9	[189]
	2020	130	1.5	0.75–2.5	1.75	[190]

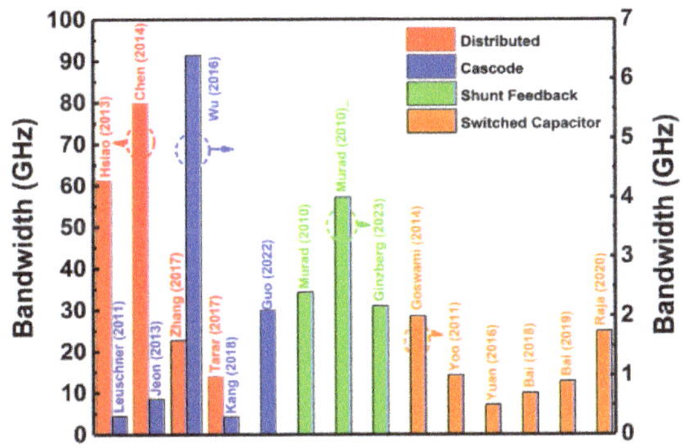

Figure 32. Frequency bandwidth achieved by the proposed bandwidth enhancement techniques [154–157,161–165,167–169,181–184,189,190].

6. Conclusions

This review paper discusses the trending state-of-the-art CMOS RFPAs in wireless communication systems. As the number of wireless devices and mobile data usage increase, there is a growing need for enhancements and upgrades to current wireless communication systems. This review provides a concise discussion of the performance metrics that are important for designing a CMOS PA, followed by an overview of recent research on CMOS PAs' performance enhancement techniques, including efficiency, linearity, and bandwidth. This review concludes that there is a trade-off between efficiency, linearity, and bandwidth in CMOS PAs, and several enhancement techniques are developed to improve their performance. The on-chip transformer and out-phasing techniques are capable of delivering watt-level maximum output power for CMOS PAs. This is due to the usage of power-combining architectures in these techniques that boost the maximum output power. It can also be deduced that CMOS PAs are still limited in delivering watt-level linear output power, since PAs are required to operate at backed-off output power to sustain the linear transmission. The switched-capacitor-based CMOS PA has the most flexibility and configurability in tuning its parameters according to its need, since its impedance can be tuned.

Author Contributions: Conceptualization, S.S.H., S.M., J.R. and B.S.Y.; methodology, S.S.H., S.M. and J.R.; software, S.M. and J.R.; formal analysis, S.M., J.R., N.K., N.A.R. and A.N.; validation, S.M., J.R., N.K., A.N. and B.S.Y.; resources, S.S.H., J.R., S.M., A.S.R. and N.K.; data curation, J.R., S.M., N.A.R. and A.N.; writing—original draft preparation, S.S.H., J.R., S.M. and N.K.; writing—review and editing, J.R., S.M., N.K. and A.N.; visualization, S.M., N.A.R., A.S.R. and N.K.; supervision, J.R. and A.N.; project administration, J.R. and S.M.; funding acquisition, J.R. and S.M. All authors have read and agreed to the published version of the manuscript.

Funding: The Malaysian Ministry of Higher Education's Fundamental Research Grant Scheme (grant number: FRGS/1/2019/TK04/USM/02/14).

Data Availability Statement: Not applicable.

Conflicts of Interest: The authors declare no conflict of interest.

References

1. Ericsson. Ericsson Mobility Report June 2016. 2016. Available online: https://www.ericsson.com/res/docs/2016/ericsson-mobility-report-2016.pdf (accessed on 12 April 2023).
2. Lie, D.Y.C. RF-SoC': Integration trends of on-chip CMOS power amplifier: Benefits of external PA versus integrated PA for portable wireless communications. *Int. J. Microw. Sci. Technol.* **2010**, 380108. [CrossRef]
3. Bennett, H.S.; Brederlow, R.; Costa, J.C.; Cottrell, P.E.; Huang, W.M.; Immorlica, A.A.; Mueller, J.E.; Racanelli, M.; Shichijo, H.; Weitzal, C.E.; et al. Device and technology evolution for Si-based RF integrated circuits. *IEEE Trans. Electron Devices Devices* **2005**, *52*, 1235–1258. [CrossRef]
4. Larson, L.E. Silicon technology tradeoffs for radio-frequency/mixed-signal 'systems-on-a-chip'. *IEEE Trans. Electron Devices* **2003**, *50*, 683–699. [CrossRef]
5. Chang, C.S.; Chao, C.P.; Chern, J.G.J.; Sun, J.Y. Advanced CMOS technology portfolio for RF IC applications. *IEEE Trans. Electron Devices* **2005**, *52*, 1324–1334. [CrossRef]
6. Abidi, A.A. RF CMOS comes of age. *IEEE J. Solid-State Circuits* **2004**, *39*, 549–561. [CrossRef]
7. Hajimiri, A. Fully integrated RF CMOS power amplifiers—A prelude to full radio integration. In Proceedings of the 2005 IEEE Radio Frequency Integrated Circuits (RFIC) Symposium—Digest of Papers, Long Beach, CA, USA, 12–14 June 2005; pp. 439–442. [CrossRef]
8. Hajimiri, A. Next-Generation CMOS RF Power Amplifiers. *IEEE Microw. Mag.* **2011**, *12*, 38–45. [CrossRef]
9. Niknejad, A.A.M.; Chowdhury, D.; Chen, J. Design of CMOS Power Amplifiers. *IEEE Trans. Microw. Theory Tech.* **2012**, *60*, 1784–1796. [CrossRef]
10. Nobbe, D.W. Silicon technology status and perspectives for multi-band and multi-standard challenges in upcoming RF frontends. In Proceedings of the 2008 IEEE Radio and Wireless Symposium, Orlando, FL, USA, 22–24 January 2008; pp. 191–194. [CrossRef]
11. Kang, D.; Park, B.; Kim, D.; Kim, J.; Cho, Y.; Kim, B. Envelope-Tracking CMOS Power Amplifier Module for LTE Applications. *IEEE Trans. Microw. Theory Tech.* **2013**, *61*, 3763–3773. [CrossRef]
12. Mariappan, S.; Rajendran, J.; Noh, N.M.; Ramiah, H. A 29dBm OIP3 Dual-Stage Power Amplifier with Analog Pre-Distorter in 0.18μm CMOS for IoT Transceiver. *IETE J. Res.* **2019**, *68*, 2298–2304. [CrossRef]
13. Mariappan, S.; Rajendran, J.; Noh, N.M.; Ramiah, H. Low voltage CMOS power amplifier with integrated analog Pre-Distorter for BLE 4.0 application. *Indonesian J. Electr. Eng. Comput.* **2019**, *14*, 895–902. [CrossRef]
14. Mariappan, S.; Rajendran, J.; Noh, N.M.; Ramiah, H.; Manaf, A.A. Energy efficiency in CMOS power amplifier designs for ultra-low power mobile wireless communication systems. *Turk. J. Electr. Eng. Comput. Sci.* **2020**, *28*, 1–16. [CrossRef]
15. Mariappan, S.; Rajendran, J.; Noh, N.M.; Ramiah, H.; Manaf, A.A. The Evolution of Integrated CMOS Power Amplifiers for Next Generation Mobile Wireless Transceivers. *J. Circuits Syst. Comput* **2019**, *29*, 2030007. [CrossRef]
16. Johansson, T.; Fritzin, J. A Review of Watt-Level CMOS RF Power Amplifiers. *IEEE Trans. Microw. Theory Tech.* **2014**, *62*, 111–124. [CrossRef]
17. Eswaran, U.; Ramiah, H.; Kanesan, J.; Reza, A.W. Energy saving power amplifier design methodologies for mobile wireless communications. *Renew. Sustain. Energy Rev.* **2015**, *51*, 1721–1727. [CrossRef]
18. Gunasegaran, P.; Rajendran, J.; Mariappan, S.; Yusof, Y.M.; Aziz, Z.A.A.; Kumar, N. A fully matched dual stage CMOS power amplifier with integrated passive linearizer attaining 23 dB gain, 40% PAE and 28 dBm OIP3. *Microelectron. Int.* **2021**, *38*, 66–77. [CrossRef]
19. Sapawi, R.; Mohamad, D.H.A.; Yusuf, D.S.A.A.; Sahari, S.K.; Salleh, D.N.S.D.A.; Hazis, N.A.H.A.; Murad, S.A.Z. CMOS Power Amplifier Design Techniques for UWB Communication: A Review. *J. Telecommun. Electron. Comput. Eng.* **2017**, *9*, 1–8.
20. Yin, Y.; Chi, B.; Xia, Z.; Wang, Z. A Reconfigurable Dual-Mode CMOS Power Amplifier with Integrated T/R Switch for 0.1–1.5 GHz Multistandard Applications. *IEEE Trans. Circuits Syst. II Express Briefs* **2014**, *61*, 471–475. [CrossRef]
21. Shim, S.; Han, J.; Hong, S. A CMOS RF Polar Transmitter of a UHF Mobile RFID Reader for High Power Efficiency. *IEEE Microw. Wirel. Compon. Lett.* **2008**, *18*, 635–637. [CrossRef]

22. Quinlan, P.; Crowley, P.; Chanca, M.; Hudson, S.; Hunt, B.; Mulvaney, K.; Reta, G.; O'Sullivan, C.E.; Walsh, P. A multimode 0.3–200-kb/s transceiver for the 433/868/915-MHz bands in 0.25-µm CMOS. *IEEE J. Solid-State Circuits* **2004**, *39*, 2297–2310. [CrossRef]
23. Osseiran, A.; Monserrat, J.F.; Marsch, P. *5G Mobile and Wireless Communications Technology*; Cambridge University Press: Cambridge, UK, 2016.
24. Walsh, K.; Johnson, J. 3G/4G Multimode Cellular Front-End Challenges; Impact on Power Amplifier Design. *Microw. J.* **2009**, *3*, 56–68.
25. *Specification of the Bluetooth System—'Bluetooth Specification Version 2.0 + EDR'*; 2004; Volume 3, p. 31. Available online: https://www.bluetooth.com/specifications/specs/core-specification-2-0-edr/ (accessed on 28 July 2023).
26. Miller, M. *Discovering Bluetooth*; SYBEX: Alameda, CA, USA, 2001.
27. Skaria, G. Class F and Inverse Class F Power Amplifier Subject to Electrical Stress Effect. Master's Thesis, University of Central Florida, Orlando, FL, USA, 2011.
28. Anakabe, A.; Ayllón, N.; Collantes, J.M.; Mallet, A.; Soubercaze-Pun, G.; Narendra, K. Automatic pole-zero identification for multivariable large-signal stability analysis of RF and microwave circuits. In Proceedings of the 40th European Microwave Conference, Paris, France, 28–30 September 2010; pp. 477–480.
29. Dellier, S.; Gourseyrol, R.; Soubercaze-Pun, G.; Collantes, J.; Anakabe, A.; Narendra, K. Stability analysis of microwave circuits. In Proceedings of the WAMICON 2012 IEEE Wireless & Microwave Technology Conference, Cocoa Beach, FL, USA, 15–17 April 2012; pp. 1–5.
30. Narendra, K.; Limiti, E.; Paoloni, C.; Collantes, J.; Jansen, R.; Yarman, S. Vectorially Combined Distributed Power Amplifiers for Software-Defined Radio Applications. *IEEE Trans. Microwave Theory Tech.* **2012**, *60*, 3189–3200. [CrossRef]
31. Imbornone, J.F.; Murphy, M.T.; Donahue, R.S.; Heaney, E. New insight into subharmonic oscillation mode of GaAs power amplifiers under severe output mismatch condition. *IEEE J. Solid-State Circuits* **1997**, *32*, 1319–1325. [CrossRef]
32. Qiao, D.; Zhao, Y.; Hung, T.; Kimball, D.; Li, M.; Asbeck, P.; Choi, D.; Kelly, D. Antenna impedance mismatch measurement and correction for adaptive CDMA transceivers. In Proceedings of the IEEE MTT-S International Microwave Symposium Digest, Long Beach, CA, USA, 17 June 2005.
33. Van Bezooijen, A.; Chanlo, C.; Van Roermund, A.H.M. Adaptively preserving power amplifier linearity under antenna mismatch. In Proceedings of the 2004 IEEE MTT-S International Microwave Symposium Digest (IEEE Cat. No. 04CH37535), Fort Worth, TX, USA, 6–11 June 2004; Volume 3, pp. 1515–1518.
34. Keerti, A.; Pham, A.H. RF Characterization of SiGe HBT Power Amplifiers Under Load Mismatch. *IEEE Trans. Microw. Theory Tech.* **2007**, *55*, 207–214. [CrossRef]
35. Karoui, W.; Parra, T. A protection circuit for HBT RF Power Amplifier under load mismatch conditions. In Proceedings of the 2008 Joint 6th International IEEE Northeast Workshop on Circuits and Systems and TAISA Conference, Montreal, QC, Canada, 22–25 June 2008; pp. 241–244. [CrossRef]
36. Suárez, A.; Ramírez, F.; Sancho, S. Stability Analysis of Power Amplifiers Under Output Mismatch Effects. *IEEE Trans. Microw. Theory Tech.* **2014**, *62*, 2273–2289. [CrossRef]
37. De Vreede, L.C.N.; van der Heijden, M.P. Linearization Techniques at the Device and Circuit Level (Invited). In Proceedings of the 2006 Bipolar/BiCMOS Circuits and Technology Meeting, Maastricht, The Netherlands, 8–10 October 2006; pp. 1–8.
38. Texas Instrument. Intermodulation Distortion (IMD). Available online: http://www.ti.com (accessed on 16 May 2023).
39. Aitchison, C.S.; Mbabele, M. Third Order Intermodulation Improvement in Distributed Amplifiers. In Proceedings of the 31st European Microwave Conference, London, UK, 24–28 September 2001; pp. 1–4. [CrossRef]
40. Minasian, R.A. Intermodulation Distortion Analysis of MESFET Amplifiers Using the Volterra Series Representation. *IEEE Trans. Microw. Theory Tech.* **1980**, *28*, 1–8. [CrossRef]
41. Asbeck, P. Linearisation of RF Power Amplifiers. Ph.D. Thesis, Nokia Mobile Phones and Department of Information Technology, Technical University of Denmark, Lyngby, Denmark, 2001.
42. Aslan, M.P. Adaptive Digital Predistortion for Power Amplifier Linearization. Master's Thesis, The Institute of Engineering and Science of Bilkent University, Ankara, Turkey, 2008.
43. El-Khatib, Z. Highly-Linearized CMOS Distributed Bidirectional Amplifier with Cross-Coupled Compensator for Wireless Communications. Ph.D. Thesis, Ottawa-Carleton Institute for Electrical Engineering, Carleton University, Ottawa, ON, Canada, 2012.
44. Liu, C.; Xiao, H.; Wu, Q.; Li, F. Spectrum design of RF power amplifier for wireless communication systems. *IEEE Trans. Consum. Electron* **2002**, *48*, 72–80. [CrossRef]
45. Boulejfen, N.; Harguem, A.; Hammi, O.; Ghannouchi, F.M.; Gharsallah, A. Analytical prediction of spectral regrowth and correlated and uncorrelated distortion in multicarrier wireless transmitters exhibiting memory effects. *IET Microw. Antennas Propag.* **2010**, *4*, 685–696. [CrossRef]
46. Zhou, G.T.; Kenney, J.S. Predicting spectral regrowth of nonlinear power amplifiers. *IEEE Trans. Commun.* **2002**, *50*, 718–722. [CrossRef]
47. Nader, C.; Händel, P.; Björsell, N. Peak-to-Average Power Reduction of OFDM Signals by Convex Optimization: Experimental Validation and Performance Optimization. *IEEE Trans. Instrum. Meas.* **2011**, *60*, 473–479. [CrossRef]

48. Ciccognani, W.; Colantonio, P.; Giannini, F.; Limiti, E.; Rossi, M. AM/AM and AM/PM power amplifier characterisation technique. In Proceedings of the 15th International Conference on Microwaves, Radar and Wireless Communications, Warsaw, Poland, 17–19 May 2004; pp. 678–681.
49. Katz, A. Linearization: Reducing distortion in power amplifiers. *IEEE Microw. Mag.* **2001**, *2*, 37–49. [CrossRef]
50. An, K.H.; Lee, O.; Kim, H.; Lee, D.H.; Han, J.; Yang, K.S.; Kim, Y.; Chang, J.J.; Woo, W.; Lee, C.-H. Power-Combining Transformer Techniques for Fully-Integrated CMOS Power Amplifiers. *IEEE J. Solid-State Circuits* **2008**, *43*, 1064–1075. [CrossRef]
51. Haldi, P.; Chowdhury, D.; Reynaert, P.; Liu, G.; Niknejad, A.M. A 5.8 GHz 1 V Linear Power Amplifier Using a Novel On-Chip Transformer Power Combiner in Standard 90 nm CMOS. *IEEE J. Solid-State Circuits* **2008**, *43*, 1054–1063. [CrossRef]
52. Chowdhury, D.; Hull, C.D.; Degani, O.B.; Wang, Y.; Niknejad, A.M. A Fully Integrated Dual-Mode Highly Linear 2.4 GHz CMOS Power Amplifier for 4G WiMax Applications. *IEEE J. Solid-State Circuits* **2009**, *44*, 3393–3402. [CrossRef]
53. Aoki, I.; Kee, S.D.; Rutledge, D.; Hajimiri, A. A 2.4-GHz, 2.2-W, 2-V fully-integrated CMOS circular-geometry active-transformer power amplifier. In Proceedings of the IEEE 2001 Custom Integrated Circuits Conference, San Diego, CA, USA, 9 May 2001; pp. 57–60. [CrossRef]
54. Aoki, I.; Kee, S.D.; Rutledge, D.B.; Hajimiri, A. Distributed active transformer-a new power-combining and impedance-transformation technique. *IEEE Trans. Microw. Theory Tech* **2002**, *50*, 316–331. [CrossRef]
55. Javidan, J.; Atarodi, M.; Luong, H.C. High Power Amplifier Based on a Transformer-Type Power Combiner in CMOS Technology. *IEEE Trans. Circuits Syst. II* **2010**, *57*, 838–842. [CrossRef]
56. Liu, G.; Liu, T.K.; Niknejad, A.M. A 1.2V, 2.4 GHz Fully Integrated Linear CMOS Power Amplifier with Efficiency Enhancement. In Proceedings of the IEEE Custom Integrated Circuits Conference 2006, San Jose, CA, USA, 10–13 September 2006; pp. 141–144. [CrossRef]
57. Liu, G.; Haldi, P.; Liu TJ, K.; Niknejad, A.M. Fully Integrated CMOS Power Amplifier with Efficiency Enhancement at Power Back-Off. *IEEE J. Solid-State Circuits* **2008**, *43*, 600–609. [CrossRef]
58. Afsahi, A.; Larson, L.E. An integrated 33.5 dBm linear 2.4 GHz power amplifier in 65nm CMOS for WLAN applications. In Proceedings of the IEEE Custom Integrated Circuits Conference 2010, San Jose, CA, USA, 19–22 September 2010; pp. 1–4. [CrossRef]
59. An, K.H.; Kim, Y.; Lee, O.; Yang, K.S.; Kim, H.; Woo, W.; Chang, J.J.; Lee, C.-H.; Kim, H.; Laskar, J. A Monolithic Voltage-Boosting Parallel-Primary Transformer Structures for Fully Integrated CMOS Power Amplifier Design. In Proceedings of the 2007 IEEE Radio Frequency Integrated Circuits (RFIC) Symposium, Honolulu, HI, USA, 3–5 June 2007; pp. 419–422. [CrossRef]
60. Belabad, A.R.; Masoumi, N.; Ashtiani, S.J. A fully integrated 2.4 GHz CMOS high power amplifier using parallel class A & B power amplifier and power-combining transformer for WiMAX application. *Int. J. Electron. Commun.* **2013**, *67*, 1030–1037. [CrossRef]
61. Ahn, H.; Baek, S.; Nam, I.; An, D.; Lee, J.K.; Jeong, M.; Kim, B.-E.; Choi, J.; Lee, O. A Fully Integrated Dual-Mode CMOS Power Amplifier with an Autotransformer-Based Parallel Combining Transformer. *IEEE Microw. Wirel. Compon. Lett.* **2017**, *27*, 833–835. [CrossRef]
62. Tsai, J.H. Design of a 5.2-GHz CMOS Power Amplifier Using TF-Based 2-Stage Dual-Radial Power Splitting/Combining Architecture. *IEEE Trans. Circuits Syst. I Regul. Pap.* **2019**, *7*, 68. [CrossRef]
63. Tsai, J.-H. Design of a 5.3-GHz 31.3-dBm Fully Integrated CMOS Power Amplifier Using Folded Splitting and Combining Architecture. *IEEE Trans. Very Large Scale Integr. Syst.* **2019**, *27*, 1527–1536. [CrossRef]
64. Choi, H.-W.; Choi, S.; Lim, J.T.; Kim, C.Y. 1-W, High-Gain, High-Efficiency, and Compact Sub-GHz Linear Power Amplifier Employing a 1:1 Transformer Balun in 180-nm CMOS. *IEEE Microw. Wirel. Compon. Lett.* **2020**, *30*, 779–781. [CrossRef]
65. Wang, T.; Qian, H.J.; Yang, B.; Luo, X. 1.2–2.8-GHz 32.4-dBm Digital Power Amplifier with Balance-Compensated Matching Network. *IEEE Microw. Wirel. Compon. Lett.* **2021**, *31*, 41–44. [CrossRef]
66. Yang, B.; Qian, H.J.; Luo, X. Quadrature Switched/Floated Capacitor Power Amplifier with Reconfigurable Self-Coupling Canceling Transformer for Deep Back-Off Efficiency Enhancement. *IEEE J. Solid-State Circuits* **2021**, *56*, 3715–3727. [CrossRef]
67. Yang, B.; Qian, H.J.; Wang, T.; Luo, X. A CMOS Wideband Watt-Level 4096-QAM Digital Power Amplifier Using Reconfigurable Power-Combining Transformer. *IEEE J. Solid-State Circuits* **2023**, *58*, 357–370. [CrossRef]
68. Doherty, W.H. A New High Efficiency Power Amplifier for Modulated Waves. *Proc. Inst. Radio Eng.* **1936**, *24*, 1163–1182. [CrossRef]
69. Vasjanov, A.; Barzdenas, V. Comparison of CMOS Power Amplifier Topologies for a Wideband Doherty Architecture. In Proceedings of the 2017 Open Conference of Electrical, Electronic and Information Sciences (eStream), Vilnius, Lithuania, 27 April 2017; pp. 1–4. [CrossRef]
70. Eswaran, U.; Ramiah, H.; Kanesan, J. Power Amplifier Design Methodologies for Next Generation Wireless Communications. *IETE Tech. Rev.* **2014**, *31*, 241–248. [CrossRef]
71. Camarchia, V.; Pirola, M.; Quaglia, R.; Jee, S.; Cho, Y.; Kim, B. The Doherty Power Amplifier: Review of Recent Solutions and Trends. *IEEE Trans. Microw. Theory Tech.* **2015**, *63*, 559–571. [CrossRef]
72. Deltimple, N.; Carneiro, M.L.; Kerhervé, E.; Carvalho, P.H.P.; Belot, D. Integrated Doherty RF CMOS Power Amplifier Design for Average Efficiency Enhancement. In Proceedings of the 2015 IEEE International Wireless Symposium (IWS 2015), Shenzhen, China, 30 March–1 April 2015; pp. 1–4. [CrossRef]

73. Hu, S.; Kousai, S.; Wang, H. A Broadband Mixed-Signal CMOS Power Amplifier with a Hybrid Class-G Doherty Efficiency Enhancement Technique. *IEEE J. Solid-State Circuits* **2016**, *51*, 598–613. [CrossRef]
74. Kaymaksut, E.; Reynaert, P. Dual-Mode CMOS Doherty LTE Power Amplifier with Symmetric Hybrid Transformer. *IEEE J. Solid-State Circuits* **2015**, *50*, 1974–1987. [CrossRef]
75. Aoki, I.; Kee, S.D.; Rutledge, D.B.; Hajimiri, A. Fully integrated CMOS power amplifier design using the distributed active-transformer architecture. *IEEE J. Solid-State Circuits* **2002**, *37*, 371–383. [CrossRef]
76. Kaymaksut, E.; Reynaert, P. Transformer-Based Uneven Doherty Power Amplifier in 90 nm CMOS for WLAN Applications. *IEEE J. Solid-State Circuits* **2012**, *47*, 1659–1671. [CrossRef]
77. Levy, C.S.; Vorapipat, V.; Buckwalter, J.F. A 14-GHz, 22-dBm Series Doherty Power Amplifier in 45-nm CMOS SOI. In Proceedings of the 2016 IEEE Compound Semiconductor Integrated Circuit Symposium (CSICS), Austin, TX, USA, 23–26 October 2016; pp. 1–4. [CrossRef]
78. Traiche, S.; Slimane, A. 3 GHz CMOS Doherty power amplifier for high efficiency. In Proceedings of the 2017 Seminar on Detection Systems Architectures and Technologies (DAT), Algiers, Algeria, 20–22 February 2017; pp. 1–4. [CrossRef]
79. Jung, H.; Zhao, J.; Wang, H. A CMOS Highly Linear Doherty Power Amplifier with Multigated Transistors. *IEEE Trans. Microw. Theory Tech.* **2019**, *67*, 1883–1891. [CrossRef]
80. Yin, Y.; Li, T.; Xiong, L.; Li, Y.; Min, H.; Yan, N.; Xu, H. A Broadband Switched-Transformer Digital Power Amplifier for Deep Back-Off Efficiency Enhancement. *IEEE J. Solid-State Circuits* **2020**, *55*, 2997–3008. [CrossRef]
81. Asbeck, P.; Popovic, Z. ET Comes of Age: Envelope Tracking for Higher-Efficiency Power Amplifiers. *IEEE Microw. Mag.* **2016**, *17*, 16–25. [CrossRef]
82. Kim, B.; Kim, J.; Kim, D.; Son, J.; Cho, Y.; Kim, J.; Park, B. Push the Envelope: Design Concepts for Envelope-Tracking Power Amplifiers. *IEEE Microw. Mag.* **2013**, *14*, 68–81. [CrossRef]
83. Ruan, X.; Wang, Y.; Jin, Q. A review of envelope tracking power supply for mobile communication systems. *CPSS Trans. Power Electron. Appl.* **2017**, *2*, 277–291. [CrossRef]
84. Ghannouchi, F.M.; Kwan, A.K.; Younes, M.; Chen, W. DSP techniques for linearity and efficiency enhancement of multi-band envelope tracking transmitters. In Proceedings of the 2014 Asia-Pacific Microwave Conference, Sendai, Japan, 4–7 November 2014; pp. 1082–1084.
85. Woo, J.; Park, S.; Kim, U.; Kwon, Y. Dynamic Stack-Controlled CMOS RF Power Amplifier for Wideband Envelope Tracking. *IEEE Trans. Microw. Theory Tech.* **2014**, *62*, 3452–3464. [CrossRef]
86. Park, B.; Kim, D.; Kim, S.; Cho, Y.; Kim, J.; Kang, D.; Jin, S.; Moon, K.; Kim, B. High-Performance CMOS Power Amplifier with Improved Envelope Tracking Supply Modulator. *IEEE Trans. Microw. Theory Tech.* **2016**, *64*, 798–809. [CrossRef]
87. Ham, J.; Bae, J.; Kim, H.; Seo, M.; Lee, H.; Hwang, K.C.; Lee, K.-Y.; Park, C.-S.; Heo, D.; Yang, Y. CMOS Power Amplifier Integrated Circuit with Dual-Mode Supply Modulator for Mobile Terminals. *IEEE Trans. Circuits Syst. I Regul. Pap.* **2016**, *63*, 157–167. [CrossRef]
88. Kahn, L.R. Single-Sideband Transmission by Envelope Elimination and Restoration. *Proc. IRE* **1952**, *40*, 803–806. [CrossRef]
89. Mamdouh, A.; Aboudina, M.; Mohieldin, A.N.; Hussien, F. CMOS A Multi-Band High-Efficiency Envelope Elimination and Restoration CMOS Power Amplifier. *Microelectron. J.* **2020**, *102*, 104823. [CrossRef]
90. Oishi, K.; Yoshida, E.; Sakai, Y.; Takauchi, H.; Kawano, Y.; Shirai, N.; Kano, H.; Kudo, M.; Murakami, T.; Tamura, T.; et al. A 1.95 GHz Fully Integrated Envelope Elimination and Restoration CMOS Power Amplifier Using Timing Alignment Technique for WCDMA and LTE. *IEEE J. Solid-State Circuits* **2014**, *49*, 2915–2924. [CrossRef]
91. Yuan, W.; Walling, J.S. A switched-capacitor controlled digital-current modulated class-E EER transmitter. In Proceedings of the 2015 IEEE 13th International New Circuits and Systems Conference (NEWCAS), Grenoble, France, 7–10 June 2015; pp. 1–4. [CrossRef]
92. Liu, X.; Zhang, H.; Mok, P.K.T.; Loung, H.C. A Multi-Loop-Controlled AC-Coupling Supply Modulator with a Mode-Switching CMOS PA in an EER System with Envelope Shaping. *IEEE J. Solid-State Circuits* **2019**, *54*, 1553–1563. [CrossRef]
93. Lee, O.; Han, J.; An, K.H.; Lee, D.H.; Lee, K.S.; Hong, S.; Lee, C.H. A Charging Acceleration Technique for Highly Efficient Cascode Class-E CMOS Power Amplifiers. *IEEE J. Solid-State Circuits* **2010**, *45*, 2184–2197. [CrossRef]
94. Raab, F. Efficiency of Outphasing RF Power-Amplifier Systems. *IEEE Trans. Commun.* **1985**, *33*, 1094–1099. [CrossRef]
95. Hung, T.P.; Choi, D.K.; Larson, L.E.; Asbeck, P.M. CMOS Outphasing Class-D Amplifier with Chireix Combiner. *IEEE Microw. Wirel. Compon. Lett.* **2007**, *17*, 619–621. [CrossRef]
96. Xu, H.; Palaskas, Y.; Ravi, A.; Sajadieh, M.; Elmala, M.; Soumyanath, K. A 28.1 dBm class-D outphasing power amplifier in 45nm LP digital CMOS. In Proceedings of the 2009 Symposium on VLSI Circuits, Kyoto, Japan, 16–18 June 2009; pp. 206–207.
97. Ding, L.; Hur, J.; Banerjee, A.; Hezar, R.; Haroun, B. A 25 dBm Outphasing Power Amplifier with Cross-Bridge Combiners. *IEEE J. Solid-State Circuits* **2015**, *50*, 1107–1116. [CrossRef]
98. Beltran, R.; Raab, F.H.; Velazquez, A. HF outphasing transmitter using class-E power amplifiers. In Proceedings of the 2009 IEEE MTT-S International Microwave Symposium Digest, Boston, MA, USA, 7–12 June 2009; pp. 757–760. [CrossRef]
99. Hu, Z.; de Vreede, L.C.N.; Alavi, M.S.; Calvillo-Cortes, D.A.; Staszewski, R.B.; He, S. A 5.9 GHz RFDAC-based outphasing power amplifier in 40-nm CMOS with 49.2% efficiency and 22.2 dBm power. In Proceedings of the 2016 IEEE Radio Frequency Integrated Circuits Symposium (RFIC), San Francisco, CA, USA, 22–24 May 2016; pp. 206–209. [CrossRef]

100. Martelius, M.; Stadius, K.; Lemberg, J.; Nieminen, T.; Roverato, E.; Kosunen, M.; Ryynänen, J.; Anttila, L.; Valkama, M. Class D CMOS power amplifier with on/off logic for a multilevel outphasing transmitter. In Proceedings of the 2016 IEEE International Symposium on Circuits and Systems (ISCAS), Montreal, QC, Canada, 22–25 May 2016; pp. 710–713. [CrossRef]
101. Banerjee, A.; Hezar, R.; Ding, L.; Haroun, B. A 29.5 dBm Class-E Outphasing RF Power Amplifier with Efficiency and Output Power Enhancement Circuits in 45 nm CMOS. *IEEE Trans. Circuits Syst. I* **2017**, *64*, 1977–1988. [CrossRef]
102. Ghahremani, A.; Annema, A.; Nauta, B. Outphasing Class-E Power Amplifiers: From Theory to Back-Off Efficiency Improvement. *IEEE J. Solid-State Circuits* **2018**, *53*, 1374–1386. [CrossRef]
103. Martelius, M.; Stadius, K.; Lemberg, J.; Roverato, E.; Nieminen, T.; Antonov, Y.; Anttila, L.; Valkama, M.; Kosunen, M.; Ryynänen, J. A Class-D Tri-Phasing CMOS Power Amplifier with an Extended Marchand-Balun Power Combiner. *IEEE Trans. Microw. Theory Tech.* **2020**, *68*, 1022–1034. [CrossRef]
104. Banerjee, A.; Ding, L.; Hezar, R. A High Efficiency Multi-Mode Outphasing RF Power Amplifier with 31.6 dBm Peak Output Power in 45nm CMOS. *IEEE Trans. Circuits Syst. I Regul. Pap.* **2020**, *67*, 815–828. [CrossRef]
105. Kang, S.; Choi, B.; Kim, B. Linearity analysis of CMOS for RF application. *IEEE Trans. Microw. Theory Tech.* **2003**, *51*, 972–977. [CrossRef]
106. Golara, S.; Moloudi, S.; Abidi, A.A. Processes of AM-PM Distortion in Large-Signal Single-FET Amplifiers. *IEEE Trans. Circuits Syst. I Regul. Pap.* **2017**, *64*, 245–260. [CrossRef]
107. Xi, T.; Huang, S.; Guo, S.; Gui, P.; Huang, D.; Chakraborty, S. High-Efficiency E-Band Power Amplifiers and Transmitter Using Gate Capacitance Linearization in a 65-nm CMOS Process. *IEEE Trans. Circuits Syst. II Express Briefs* **2017**, *64*, 234–238. [CrossRef]
108. Wang, C.; Vaidyanathan, M.; Larson, L.E. A capacitance-compensation technique for improved linearity in CMOS class-AB power amplifiers. *IEEE J. Solid-State Circuits* **2004**, *39*, 1927–1937. [CrossRef]
109. Camarchia, V.; Colantonio, P.; Giannini, F.; Giofrè, R.; Jiang, T.; Pirola, M.; Quaglia, R.; Ramella, C. A Design Strategy for AM/PM Compensation in GaN Doherty Power Amplifiers. *IEEE Access* **2017**, *5*, 22244–22251. [CrossRef]
110. Cho, M. Analog Predistortion for Improvement of Rf Power Amplifier Efficiency and Linearity. Ph.D. Thesis, Georgia Institute of Technology, Atlanta, GA, USA, 2016.
111. Li, F. Linearization of Power Amplifiers in Wide Band Communication Systems by Digital Baseband Predistortion Technique. Ph.D. Thesis, Universite De Nantes, Nantes, France, 2012.
112. Borel, A.; Barzdėnas, V. A Review of the Basic Power Amplifier Linearization Methods. In Proceedings of the 2020 IEEE Open Conference of Electrical, Electronic and Information Sciences (eStream), Vilnius, Lithuania, 30 April 2020; pp. 1–4. [CrossRef]
113. Dawson, J.L.; Lee, T.H. Cartesian feedback for RF power amplifier linearization. In Proceedings of the 2004 American Control Conference, Boston, MA, USA, 30 June–2 July 2004; Volume 1, pp. 361–366. [CrossRef]
114. Monticelli, D. System approaches to power management. In Proceedings of the APEC Seventeenth Annual IEEE Applied Power Electronics Conference and Exposition (Cat. No.02CH37335), Dallas, TX, USA, 10–14 March 2002; Volume 1, pp. 3–7. [CrossRef]
115. Tee, L.; Sacchi, E.; Bocock, R.; Wongkomet, N.; Gray, P. A Cartesian-Feedback Linearized CMOS RF Transmitter for EDGE Modulation. In Proceedings of the 2006 Symposium on VLSI Circuits, 2006, Digest of Technical Papers, Honolulu, HI, USA, 15–17 June 2006; pp. 232–233. [CrossRef]
116. Atress, A.B.; Raafat, H.; Shawkey, H.; Zaki, A. Zigbee Power Amplifier Linearization Using Cartesian Feedback. *Int. J. Adv. Res.* **2016**, *4*, 1769–1776.
117. Sowlati, T.; Rozenblit, D.; Pullela, R.; Damgaard, M.; McCarthy, E.; Koh, D.; Ripley, D.; Balteanu, F.; Gheorghe, I. Quad-band GSM/GPRS/EDGE polar loop transmitter. *IEEE J. Solid-State Circuits* **2004**, *39*, 2179–2189. [CrossRef]
118. Nakatani, T.; Kimball, D.; Larson, L.; Asbeck, P. 0.7–1.8 GHz multiband digital polar transmitter using watt-class current-mode class-D CMOS power amplifier and digital envelope modulation technique for reduced spurious emissions. *Int. J. Microwave Wireless Technol.* **2013**, *5*, 271–284. [CrossRef]
119. Zheng, S.; Luong, H.C. A CMOS WCDMA/WLAN Digital Polar Transmitter with AM Replica Feedback Linearization. *IEEE J. Solid-State Circuits* **2013**, *48*, 1701–1709. [CrossRef]
120. Liao, H.; Chen, J.; Chiou, H.; Chen, C. High-linearity CMOS feedforward power amplifier for WiMAX application. In Proceedings of the 2008 Asia-Pacific Microwave Conference, Hong Kong, China, 16–20 December 2008; pp. 1–4. [CrossRef]
121. Chen, Y.-H.; Kao, K.; Chao, C.-Y.; Lin, K. A 24 GHz CMOS power amplifier with successive IM2 feed-forward IMD3 cancellation. In Proceedings of the 2015 IEEE MTT-S International Microwave Symposium, Phoenix, AZ, USA, 17–22 May 2015; pp. 1–4. [CrossRef]
122. Garcia, P.; de Mingo, J.; Valdovinos, A.; Ortega, A. An adaptive digital method of imbalances cancellation in LINC transmitters. *IEEE Trans. Veh. Technol.* **2005**, *54*, 879–888. [CrossRef]
123. Lee, H.; Lee, H.; Paek, J.; Hong, S. A CMOS power amplifier for Multi-mode LINC architecture. In Proceedings of the 2010 IEEE Radio and Wireless Symposium (RWS), New Orleans, LA, USA, 10–14 January 2010; pp. 41–44. [CrossRef]
124. Lee, H.; Jang, S.; Hong, S. A Hybrid Polar-LINC CMOS Power Amplifier with Transmission Line Transformer Combiner. *IEEE Trans. Microwave Theory Tech.* **2013**, *61*, 1261–1271. [CrossRef]
125. Yao, J.; Long, S.I. Power Amplifier Selection for LINC Applications. *IEEE Trans. Circuits Syst. II Express Briefs* **2006**, *53*, 763–767. [CrossRef]
126. Katz, A.; Wood, J.; Chokola, D. The Evolution of PA Linearization: From Classic Feedforward and Feedback Through Analog and Digital Predistortion. *IEEE Microw. Mag.* **2016**, *17*, 32–40. [CrossRef]

127. Rey, C.G. Predistorter linearizes CDMA power amplifiers. *Microwaves RF Mag.* **1998**, *44*, 114–123.
128. Yi, J.; Yang, Y.; Park, M.; Kang, W.; Kim, B. Analog predistortion linearizer for high-power RF amplifiers. *IEEE Trans. Microw. Theory Tech.* **2000**, *48*, 2709–2713. [CrossRef]
129. Son, K.Y.; Koo, B.; Hong, S. A CMOS Power Amplifier with a Built-In RF Predistorter for Handset Applications. *IEEE Trans. Microwave Theory Tech.* **2012**, *60*, 2571–2580. [CrossRef]
130. Paramesh, J.; Bishop, R.; Soumyanath, K.; Allstot, D.J. A four-antenna receiver in 90-nm CMOS for beamforming and spatial diversity. *IEEE J. Solid-State Circuits* **2005**, *40*, 2515–2524. [CrossRef]
131. Afsahi, A.; Rael, J.J.; Behzad, A.; Chien, H.M.; Pan, M.; Au, S.; Ojo, A.; Lee, C.P.; Anand, S.B.; Chien, K.; et al. A Low-Power Single-Weight-Combiner 802.11abg SoC in 0.13 µm CMOS for Embedded Applications Utilizing An Area and Power Efficient Cartesian Phase Shifter and Mixer Circuit. *IEEE J. Solid-State Circuits* **2008**, *43*, 1101–1118. [CrossRef]
132. Rostomyan, N.; Jayamon, J.A.; Asbeck, P.M. 15 GHz Doherty Power Amplifier with RF Predistortion Linearizer in CMOS SOI. *IEEE Trans. Microw. Theory Techn.* **2018**, *66*, 1339–1348. [CrossRef]
133. Mariappan, S.; Rajendran, J.; Ramiah, H.; Mak, P.-I.; Yin, J.; Martins, R.P. An 800 MHz-to-3.3 GHz 20-MHz Channel Bandwidth WPD CMOS Power Amplifier For Multiband Uplink Radio Transceivers. *IEEE Trans. Circuits Syst. II Exp. Briefs* **2021**, *68*, 1178–1182. [CrossRef]
134. Mariappan, S.; Rajendran, J.; Yusof, Y.M.; Noh, N.M.; Yarman, B.S. An 0.4–2.8 GHz CMOS Power Amplifier with On-Chip Broadband-Pre-Distorter (BPD) Achieving 36.1–38.6% PAE and 21 dBm Maximum Linear Output Power. *IEEE Access* **2021**, *9*, 48831–48840. [CrossRef]
135. Mariappan, S.; Rajendran, J.; Chen, Y.; Mak, P.-I.; Martins, R.P. A 1.7-to-2.7GHz 35–38% PAE Multiband CMOS Power Amplifier Employing a Digitally-Assisted Analog Pre-Distorter (DAAPD) Reconfigurable Linearization Technique. *IEEE Trans. Circuits Syst. II Exp. Briefs* **2021**, *68*, 3381–3385. [CrossRef]
136. Mariappan, S.; Rajendran, J.; Kumar, N.; Othman, M.; Nathan, A.; Grebennikov, A.; Yarman, B.S. A Wide-Bandwidth PVT-Reconfigurable CMOS Power Amplifier with an Integrated Tunable-Output Impedance Matching Network. *Micromachines* **2023**, *14*, 530. [CrossRef]
137. Tsai, J.-H. A 5.3-GHz 30.1-dBm Fully Integrated CMOS Power Amplifier with High-Power Built-In Linearizer. *IEEE Microw. Wirel. Technol. Lett.* **2023**, *33*, 431–434. [CrossRef]
138. Woo, W.; Kenney, J.S. A predistortion linearization system for high power amplifiers with low-frequency envelope memory effects. In Proceedings of the IEEE MTT-S International Microwave Symposium Digest, Long Beach, CA, USA, 17 June 2005; p. 4. [CrossRef]
139. Paaso, H.; Mammela, A. Comparison of direct learning and indirect learning predistortion architectures. In Proceedings of the 2008 IEEE International Symposium on Wireless Communication Systems, Reykjavik, Iceland, 21–24 October 2008; pp. 309–313. [CrossRef]
140. Zhou, D.; DeBrunner, V. A novel adaptive nonlinear predistorter based on the direct learning algorithm. In Proceedings of the 2004 IEEE International Conference on Communications (IEEE Cat. No. 04CH37577), Paris, France, 20–24 June 2004; Volume 4, pp. 2362–2366. [CrossRef]
141. Li, Y.; Zhang, X. Adaptive digital predistortion based on MC-FQRD-RLS algorithm using indirect learning architecture. In Proceedings of the 2010 2nd International Conference on Advanced Computer Control, Shenyang, China, 27–29 March 2010; pp. 240–242. [CrossRef]
142. Cho, Y.; Lee, J.; Jin, S.; Park, B.; Moon, J.; Kim, J.; Kim, B. Fully Integrated CMOS Saturated Power Amplifier with Simple Digital Predistortion. *IEEE Microw. Wirel. Compon. Lett.* **2014**, *24*, 533–535. [CrossRef]
143. Wong, A.; Godoy, P.; Carnu, O.; Li, H.; Zhao, X.; Olyaei, A.; Ghaffari, A.; Tam, S.-W.; Winoto, R.; Tsang, R. A dual-core power combining digital power amplifier for 802.11 b/g/n with +26.8 dBm linear output power in 28 nm CMOS. In Proceedings of the 2017 IEEE Radio Frequency Integrated Circuits Symposium (RFIC), Honolulu, HI, USA, 4–6 June 2017; pp. 192–195. [CrossRef]
144. Kim, W.; Yang, K.S.; Han, J.; Chang, J.J.; Lee, C.H. An EDGE/GSM Quad-Band CMOS Power Amplifier. *IEEE J. Solid-State Circuits* **2014**, *49*, 2141–2149. [CrossRef]
145. Kim, H.; Bae, J.; Ham, J.; Gu, J.; Seo, M.; Hwang, K.; Lee, K.-Y.; Park, C.-S.; Yang, Y. Efficiency enhanced CMOS digitally controlled dynamic bias switching power amplifier for LTE. *Microw. Opt. Technol. Lett.* **2015**, *57*, 2315–2321. [CrossRef]
146. Kim, B.; Lee, D.-H.; Hong, S.; Park, M. A Multi-Band CMOS Power Amplifier Using Reconfigurable Adaptive Power Cell Technique. *IEEE Microw. Wireless Compon. Lett.* **2016**, *26*, 616–618. [CrossRef]
147. Ren, J.; Dai, R.; He, J.; Zhang, Z.; Zou, S. A high power CMOS cascode power amplifier with adaptive dynamic bias control for linearity enhancement. *Microw. Opt. Technol. Lett.* **2020**, *62*, 2451–2457. [CrossRef]
148. Mariappan, S.; Rajendran, J.; Noh, N.; Yusof, Y.; Kumar, N. A 23.3 dBm CMOS power amplifier with third-order gm cancellation linearization technique achieving OIP3 of 34 dBm. *Circuit World* **2021**, *48*, 215–222. [CrossRef]
149. Rawat, A.S.; Rajendran, J.; Mariappan, S.; Shasidharan, P.; Kumar, N.; Yarman, B.S. A 919 MHz—923 MHz, 21 dBm CMOS Power Amplifier with Bias Modulation Linearization Technique Achieving PAE of 29% for LoRa Application. *IEEE Access* **2022**, *10*, 79365–79378. [CrossRef]
150. Rawat, A.S.; Rajendran, J.; Mariappan, S.; Kumar, N.; Mak, P.-I.; Martins, R.P. A 27-dBm, 0.92-GHz CMOS Power Amplifier with Mode Switching and a High-Q Compact Inductor (HQCI) Achieving a 30% Back-Off PAE. *IEEE Trans. Circuits Syst. II* **2023**, *70*, 121–125. [CrossRef]

151. Kang, S.; Sung, E.; Hong, S. Dynamic Feedback Linearizer of RF CMOS Power Amplifier. *IEEE Microw. Wireless Compon. Lett.* **2018**, *28*, 915–917. [CrossRef]
152. Eriksson, K.; Darwazeh, I.; Zirath, H. InP DHBT Distributed Amplifiers with Up to 235-GHz Bandwidth. *IEEE Trans. Microw. Theory Tech.* **2015**, *63*, 1334–1341. [CrossRef]
153. Krishnamurthy, K. Ultra-Broadband, Efficient, Microwave Power Amplifiers in Gallium Nitride HEMT Technology. Ph.D. Thesis, University of California, Santa Barbara, CA, USA, 2000.
154. Hsiao, C.; Su, T.; Hsu, S.S.H. CMOS Distributed Amplifiers Using Gate–Drain Transformer Feedback Technique. *IEEE Trans. Microw. Theory Tech.* **2013**, *61*, 2901–2910. [CrossRef]
155. Chen, P.-H.; Yeh, K.-S.; Kao, J.-C.; Wang, H. A high-performance DC-80 GHz distributed amplifier in 40-nm CMOS digital process. In Proceedings of the 2014 IEEE MTT-S International Microwave Symposium (IMS2014), Tampa, FL, USA, 1–6 June 2014; pp. 1–3. [CrossRef]
156. Zhang, Y.; Ma, K. A 2–22 GHz CMOS Distributed Power Amplifier with Combined Artificial Transmission Lines. *IEEE Microw. Wireless Compon. Lett.* **2017**, *27*, 1122–1124. [CrossRef]
157. Tarar, M.M.; Negra, R. Design and Implementation of Wideband Stacked Distributed Power Amplifier in 0.13 μm CMOS Using Uniform Distributed Topology. *IEEE Trans. Microw. Theory Tech.* **2017**, *65*, 5212–5222. [CrossRef]
158. Sira, D.; Larsen, T. Process, Voltage and Temperature Compensation Technique for Cascode Modulated PAs. *IEEE Trans. Circuits Syst. I* **2013**, *60*, 2511–2520. [CrossRef]
159. Krishnanji, S. Design of a Variable Gain Amplifier for an Ultra Wideband Receiver. Master's Thesis, Texas A&M University, College Station, TX, USA, 2005.
160. Albasha, L.; Loo, C.; Nguyen, D.; Doherty, G.; El-Gamal, M.N. An Ultra-Wideband Digitally Programmable Power Amplifier with Efficiency Enhancement for Cellular and Emerging Wireless Communication Standards. *IEEE Trans. Circuits Syst. I* **2016**, *63*, 1579–1591. [CrossRef]
161. Leuschner, S.; Mueller, J.; Klar, H. A 1.8GHz wide-band stacked-cascode CMOS power amplifier for WCDMA applications in 65nm standard CMOS. In Proceedings of the 2011 IEEE Radio Frequency Integrated Circuits Symposium, Baltimore, MD, USA, 5–7 June 2011; pp. 1–4. [CrossRef]
162. Jeon, H.; Ryu, M.J.; Kim, B.; Kim, D. A Cascode Feedback Bias Technique for Linear CMOS Power Amplifiers in a Multistage Cascode Topology. *IEEE Trans. Microw. Theory Tech.* **2013**, *61*, 890–901. [CrossRef]
163. Wu, H.; Cheng, Q.; Li, X.; Fu, H. Analysis and Design of an Ultrabroadband Stacked Power Amplifier in CMOS Technology. *IEEE Trans. Circuits Syst. II* **2016**, *63*, 49–53. [CrossRef]
164. Kang, S.; Jeong, G.; Hong, S. Study on Dynamic Body Bias Controls of RF CMOS Cascode Power Amplifier. *IEEE Microw. Wireless Compon. Lett.* **2018**, *28*, 705–707. [CrossRef]
165. Guo, T.; Lu, Z.; Tang, K.; Yang, C.; Yu, B.; Zheng, Y. A Floating-Body Transistor-Based Power Amplifier for Sub-6-GHz 5G Applications in SOI CMOS 130-nm Process. *IEEE Trans. Circuits Syst. II* **2022**, *69*, 4088–4092. [CrossRef]
166. Jose, S.; Lee, H.-J.; Ha, D.; Choi, S.S. A low-power CMOS power amplifier for ultra wideband (UWB) applications. In Proceedings of the 2005 IEEE International Symposium on Circuits and Systems, Kobe, Japan, 23–26 May 2005; Volume 5, pp. 5111–5114. [CrossRef]
167. Murad, S.A.Z.; Pokharel, R.K.; Sapawi, R.; Kanaya, H.; Yoshida, K. High efficiency, good linearity, and excellent phase linearity of 3.1–4.8 GHz CMOS UWB PA with a current-reused technique. *IEEE Trans. Consumer Electron.* **2010**, *56*, 1241–1246. [CrossRef]
168. Murad, S.A.Z.; Pokharel, R.K.; Galal, A.I.A.; Sapawi, R.; Kanaya, H.; Yoshida, K. An Excellent Gain Flatness 3.0–7.0 GHz CMOS PA for UWB Applications. *IEEE Microw. Wirel. Compon. Lett.* **2010**, *20*, 510–512. [CrossRef]
169. Ginzberg, N.; Cohen, E. A Wideband CMOS Power Amplifier with 52% Peak PAE Employing Resistive Shunt Feedback for Sub-6 GHz 5G Applications. *IEEE Microw. Wirel. Technol. Lett.* **2023**, *33*, 192–195. [CrossRef]
170. Peng, X.-Q.; Yu, W.-H.; Mak, P.-I.; Martins, R.P. A 26.3 dBm 2.5 to 6 GHz wideband class-D switched-capacitor power amplifier with 40% peak PAE. In Proceedings of the 2014 IEEE International Conference on Electron Devices and Solid-State Circuits, Chengdu, China, 1–2 June 2014; pp. 1–2. [CrossRef]
171. Aparin, V.; Dunworth, J.; Seward, L.; Yuan, W.; Walling, J.S. A transformer combined quadrature switched capacitor power amplifier in 65nm CMOS. In Proceedings of the 2016 14th IEEE International New Circuits and Systems Conference (NEWCAS), Vancouver, BC, Canada, 26–29 June 2016; pp. 1–4. [CrossRef]
172. Yuan, W.; Aparin, V.; Dunworth, J.; Seward, L.; Walling, J.S. A quadrature switched capacitor power amplifier in 65nm CMOS. In Proceedings of the 2015 IEEE Radio Frequency Integrated Circuits Symposium (RFIC), Phoenix, AZ, USA, 17–19 May 2015; pp. 135–138. [CrossRef]
173. Yuan, W.; Walling, J.S. A multiphase switched capacitor power amplifier in 130nm CMOS. In Proceedings of the 2016 IEEE Radio Frequency Integrated Circuits Symposium (RFIC), San Francisco, CA, USA, 22–24 May 2016; pp. 210–213. [CrossRef]
174. Cho, H.; Lee, M.; Park, C.; Park, J.Y. A Fully Integrated Switched Capacitor using Low Temperature and Wet Release Process for Reconfigurable CMOS Triple-band Power Amplifier. In Proceedings of the 2019 IEEE Asia-Pacific Microwave Conference (APMC), Singapore, 10–13 December 2019; pp. 1467–1469. [CrossRef]
175. Bai, Z.; Yuan, W.; Azam, A.; Walling, J.S. A split-array, C-2C switched-capacitor power amplifier in 65 nm CMOS. In Proceedings of the 2017 IEEE Radio Frequency Integrated Circuits Symposium (RFIC), Honolulu, HI, USA, 4–6 June 2017; pp. 336–339. [CrossRef]

176. Huang, C.; Chen, Y.; Zhang, T.; Sathe, V.; Rudell, J.C. A 40 nm CMOS single-ended switch-capacitor harmonic-rejection power amplifier for ZigBee applications. In Proceedings of the 2016 IEEE Radio Frequency Integrated Circuits Symposium (RFIC), San Francisco, CA, USA, 22–24 May 2016; pp. 214–217. [CrossRef]
177. Yoo, S.; Walling, J.S.; Woo, E.; Allstot, D.J. A power-combined switched-capacitor power amplifier in 90nm CMOS. In Proceedings of the 2011 IEEE Radio Frequency Integrated Circuits Symposium (RFIC), Baltimore, MD, USA, 5–7 June 2011; pp. 1–4. [CrossRef]
178. Yoo, S.; Hung, S.; Yoo, S. A Watt-Level Quadrature Class-G Switched-Capacitor Power Amplifier with Linearization Techniques. *IEEE J. Solid-State Circuits* **2019**, *54*, 1274–1287. [CrossRef]
179. Ginzberg, N.; Cohen, E. Switched-Capacitor RF Power Amplifiers: A Review of Efficiency and Linearity Considerations. In Proceedings of the 2018 IEEE International Conference on the Science of Electrical Engineering in Israel (ICSEE), Eilat, Israel, 21–23 November 2018; pp. 1–5. [CrossRef]
180. Truppi, A.; Samori, C.; Lacaita, A.L.; Levantino, S.; Ronchi, M.; Sosio, M. Impact of CMOS Scaling on Switched-Capacitor Power Amplifiers. In Proceedings of the 2018 IEEE International Symposium on Circuits and Systems (ISCAS), Florence, Italy, 27–30 May 2018; pp. 1–4. [CrossRef]
181. Goswami, S.; Kim, H.; Dawson, J.L. A Frequency-Agile RF Frontend Architecture for Multi-Band TDD Applications. *IEEE J. Solid-State Circuits* **2014**, *49*, 2127–2140. [CrossRef]
182. Yoo, S.; Walling, J.S.; Woo, E.C.; Allstot, D.J. A switched-capacitor power amplifier for EER/polar transmitters. In Proceedings of the 2011 IEEE International Solid-State Circuits Conference, San Francisco, CA, USA, 20–24 February 2011; pp. 428–430. [CrossRef]
183. Yuan, W.; Aparin, V.; Dunworth, J.; Seward, L.; Walling, J.S. A Quadrature Switched Capacitor Power Amplifier. *IEEE J. Solid-State Circuits* **2016**, *51*, 1200–1209. [CrossRef]
184. Bai, Z.; Azam, A.; Johnson, D.; Yuan, W.; Walling, J.S. Split-Array, C-2C Switched-Capacitor Power Amplifiers. *IEEE J. Solid-State Circuits*. **2018**, *53*, 1666–1677. [CrossRef]
185. Jin, H.; Kim, D.; Kim, B. 9.5 efficient digital quadrature transmitter based on IQ cell sharing. In Proceedings of the 2015 IEEE International Solid-State Circuits Conference-(ISSCC) Digest of Technical Papers, San Francisco, CA, USA, 22–26 February 2015; pp. 1–3. [CrossRef]
186. Singh, S.P.; Prabhakar, A.; Bhattacharyya, A.B. C-2C ladder voltage dividers for application in all-MOS A/D convertors. *Electron. Lett.* **1982**, *18*, 537–538. [CrossRef]
187. Lee, S.W.; Chung, H.J.; Han, C.H. C-2C digital-to-analogue converter on insulator. *Electron. Lett.* **1999**, *35*, 1242–1243. [CrossRef]
188. Xiong, W.; Guo, Y.; Zschieschang, U.; Klauk, H.; Murmann, B. A 3-V, 6-Bit C-2C Digital-to-Analog Converter Using Complementary Organic Thin-Film Transistors on Glass. *IEEE J. Solid-State Circuits* **2010**, *45*, 1380–1388. [CrossRef]
189. Bai, Z.; Azam, A.; Walling, J.S. A Frequency Tuneable Switched-Capacitor PA in 65nm CMOS. In Proceedings of the 2019 IEEE Radio Frequency Integrated Circuits Symposium (RFIC), Boston, MA, USA, 2–4 June 2019; pp. 295–298. [CrossRef]
190. Raja, I.; Banerjee, G. A 0.75–2.5-GHz All-Digital RF Transmitter with Integrated Class-E Power Amplifier for Spectrum Sharing Applications in 5G Radios. *IEEE Trans. VLSI Syst.* **2020**, *28*, 2109–2121. [CrossRef]

micromachines

MDPI

Article

Analysis and Design of a Non-Magnetic Bulk CMOS Passive Circulator Using 25% Duty-Cycle Clock

Jian Gao [1], Xinghua Wang [1], Fang Han [1], Jiayue Wan [1,*] and Wei Gu [2,*]

[1] School of Integrated Circuits and Electronics, Beijing Institute of Technology, Beijing 100081, China
[2] School of Information and Electronics, Beijing Institute of Technology, Beijing 100081, China
* Correspondence: wanjiayue@bit.edu.cn (J.W.); 6120220194@bit.edu.cn (W.G.)

Abstract: A circulator, which is a non-reciprocal device, is widely used in full-duplex systems, future communication and sensing networks, and quantum computing, and it is difficult to implement a passive topology on a chip. Based on switch-based spatio-temporal conductivity modulation, in this study, we design and implement a non-magnetic on-chip passive circulator operating at the Ku band in a 90-nm bulk CMOS technology using a 25% duty-cycle I/Q clock signal. With the virtue of the four-phase non-overlapping clock signal, the proposed circulator achieves a 3.9 dB transmitter (TX)-to-antenna (ANT) and a 4.0 dB ANT-to-receiver (RX) insertion loss with a 1-dB bandwidth of 2.7 GHz (21.4%). The TX-to-RX isolation is better than 17.2 dB, and the TX-to-ANT IIP3 and ANT-to-RX IIP3 are 19.7 dBm and 20.0 dBm, respectively, while occupying a die area of 1.55 mm × 1.15 mm. Although low-cost bulk CMOS technology is used, competitive isolation, linearity performance, and isolation bandwidth are achieved in the proposed design.

Keywords: CMOS; circulator; 25% duty cycle; full-duplex; gyrator

Citation: Gao, J.; Wang, X.; Han, F.; Wan, J.; Gu, W. Analysis and Design of a Non-Magnetic Bulk CMOS Passive Circulator Using 25% Duty-Cycle Clock. *Micromachines* **2023**, *14*, 33. https://doi.org/10.3390/mi14010033

Academic Editor: Niall Tait

Received: 12 November 2022
Revised: 18 December 2022
Accepted: 20 December 2022
Published: 23 December 2022

1. Introduction

Full-duplex (FD) communication, considered as one of the most important technologies in 5G as well as future communication and sensing systems, has been widely studied in recent years [1–7], as it allows simultaneous transmission and reception at the same time. Compared to Frequency Division Duplexing (FDD) and Time Division Duplexing (TDD) systems that use different frequency channels or time slots for transmission and reception, FD may increase the transmission speed by up to two times and save the frequency bands. However, the most significant obstacle for FD to be commercially used is the large self-interference (SI) from transmitter to receiver [3]. Therefore, a number of self-interference cancellation (SIC) technologies have been proposed in terms of the antenna interface, radio frequency (RF), and analog and digital domains [8–18] in order to mitigate this challenge.

At the antenna interface, SIC can be performed by separated antennas [7], which occupy a large area and are not suitable for phased-array or multiple input, multiple output (MIMO) systems. In order to reduce the area, a shared antenna is suggested. Typically, an electrical-balance duplexer (EBD) [19,20] is utilized to suppress SI. However, the EBD, which is a reciprocal three-port device, exhibits an inherent 3-dB loss, which is not acceptable in low-loss systems nowadays. Moreover, another candidate is a circulator, which is a non-reciprocal three-port device. Traditionally, a circulator is implemented using ferrite materials, which are bulky, expensive, and incompatible with the integrated circuit (IC) fabrication process. On-chip active circulators [21–25] using transistors suffer from poor noise and linearity performance and a limited dynamic range [26]. Therefore, a non-magnetic non-reciprocal passive on-chip circulator is expected. Nevertheless, the implementation of non-magnetic non-reciprocity is a great challenge, which has been widely studied in the past few years [27–35].

An on-chip CMOS non-magnetic passive circulator was first demonstrated by Prof. Krishnaswamy's group using an N-path filter [32,36]. Later, the first on-chip non-magnetic passive circulator based on switch-based spatio-temporal conductivity modulation [33] was proposed [37], which consumes more area. Compared to the circulator using the N-path filter, a higher operating frequency and lower clock frequency were achieved. It is also utilized in quantum computing applications [38,39] at cryogenic temperatures. A summary of the previous literature is shown in Table 1.

Table 1. Summary of prior literature and proposed work.

Reference	Topology	Technology	Frequency (GHz)	Clock Duty Cycle	Usage
[36]	N-path	65 nm	0.6–0.8	8	Sub-GHz
[40]		65 nm	0.11–1.1/ 0.28–1.15	4/8	Sub-GHz
[41]		65 nm	DC-1	8	Sub-GHz
[37]	Switched TL	45 nm SOI *	22.7–27.3	2	5G
[42]		180 nm SOI	0.86–1.08	2	Sub-GHz
[43]		45 nm SOI	50–56.8	2	WiFi
[44]		180 nm SOI	0.914–1.086	2	Sub-GHz
[45]		28 nm FDSOI **	0.05–7	2	Sub-GHz

* SOI: Silicon-on-insulator; ** FDSOI: Fully depleted silicon-on-insulator.

Nowadays, the circulator based on the N-path filter is attempted to be used in complete full-duplex transceivers [36] due to its small form factor and lower insertion loss. However, the circulator based on the N-path filter limits the transceivers to operate at lower frequencies, which is not suitable for millimeter-wave wireless systems. Traditionally, a 50% duty-cycle clock signal is applied on the circulator based on switch-based spatio-temporal conductivity modulation. However, using a 25% duty-cycle clock signal can potentially provide a higher conversion gain and lower crosstalk compared with using a 50% duty-cycle clock signal. Moreover, the design of two local oscillator (LO) signals at millimeter-wave frequencies with different duty cycles is a great challenge.

According to the previous discussion, a passive circulator operating at 10–20 GHz remains to be studied. In this frequency band, Ku-band satellite communication is located. Traditionally, the simultaneous transmission and reception of a satellite communication system is implemented by utilizing different frequency channels. However, with the increasing data volume and the emerging future communication methods, full-duplex satellite communication becomes a promising means to potentially double the throughput and save the frequency bands. Therefore, an on-chip passive circulator for full-duplex satellite communication is an emerging research topic in the near future.

To address the research gap and potentially use the 25%-duty-cycle clock signal to implement higher conversion gains, this paper describes the analysis, design, and implementation of a non-magnetic passive circulator operating at the Ku band based on switch-based spatio-temporal conductivity modulation in a 90-nm bulk CMOS technology. Section 2 begins with an illustration of the working state of the proposed circulator based on time-domain and frequency-domain analysis. A novel system-level analysis method is also presented in this section. Then, the implementation details of the proposed circulator are provided in Section 3. Sections 4 and 5 presents the measurement results and the limitation of the study and recommendations, and the paper concludes in Section 6.

2. Analysis

A gyrator [46], considered as a passive, non-reciprocal, and two-port device, was proposed in 1948, which exhibits a phase difference of 180° between the forward and reverse directions and is a good candidate for passive non-reciprocal element design, e.g., circulators. The gyrator using a 25% duty-cycle clock signal embedded in the proposed circulator is shown in Figure 1. In order to clarify the working state of the gyrator, time-domain analysis, frequency-domain analysis, and the novel system-level analysis are presented in this section.

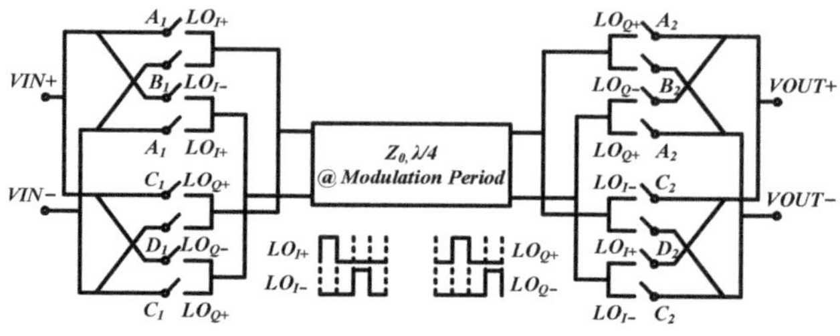

Figure 1. Gyrator using 25% duty-cycle clock signal.

2.1. Time-Domain Analysis

In order to verify the working state of the gyrator using a 25% duty-cycle clock signal, the phase shift of different modulation periods both in forward and reverse directions is illustrated, as shown in Tables 2 and 3.

Table 2. Phase shift in forward direction.

Modulation Period	Left	TL *	Right	Phase Shift
0	A_1 (0°)	90°	A_2 (0°)	90°
$T_m/4$	C_1 (0°)	90°	C_2 (0°)	90°
$T_m/2$	B_1 (180°)	90°	B_2 (180°)	450° (90°)
$3T_m/4$	D_1 (180°)	90°	D_2 (180°)	450° (90°)

* TL = The λ/4 transmission line in the gyrator.

Table 3. Phase shift in reverse direction.

Modulation Period	Right	TL *	Left	Phase Shift
0	D_2 (180°)	90°	C_1 (0°)	270° (−90°)
$T_m/4$	A_2 (0°)	90°	B_1 (180°)	270° (−90°)
$T_m/2$	C_2 (0°)	90°	D_1 (180°)	270° (−90°)
$3T_m/4$	B_2 (180°)	90°	A_1 (0°)	270° (−90°)

* TL = The λ/4 transmission line in the gyrator.

From Tables 2 and 3, it is found that the gyrator using the 25% duty-cycle clock signal exhibits a 90° phase shift in the forward direction and −90° in the reverse direction, which implements a phase difference of 180°.

2.2. Frequency-Domain Analysis

In this section, the frequency-domain analysis of the gyrator embedded in the proposed circulator is explained. Due to the double-balanced differential implementation of the switches, the final derivation can be multiplied by 2 directly. The Fourier series of the 25% duty-cycle clock signal is as follows:

$$LO(t) = \frac{2}{\pi}\sum_{n=1}^{\infty}\frac{\sin\frac{(2n-1)\pi}{4}}{2n-1}\left(e^{j(2n-1)\omega_m t} + e^{-j(2n-1)\omega_m t}\right) \quad (1)$$

For an input signal $v_1^+(t) = e^{j\omega t}$, the output signal of the gyrator at the right-hand side can be derived as

$$
\begin{aligned}
v_2^-(t) &= v_1^+\left(t-\tfrac{T_m}{4}\right) \times LO\left(t-\tfrac{T_m}{4}\right) \times LO\left(t-\tfrac{T_m}{4}\right) \\
&= \tfrac{8}{\pi^2} e^{-j\omega \frac{T_m}{4}} e^{j\omega t} \sum_{n=1}^{\infty} \frac{\sin^2\left(\frac{2n-1}{4}\right)}{(2n-1)^2} \\
&= \tfrac{1}{2} \cdot \tfrac{8}{\pi^2} e^{-j\omega \frac{T_m}{4}} e^{j\omega t} \sum_{n=1}^{\infty} \frac{1}{(2n-1)^2} \\
&= \tfrac{1}{2} e^{-j\omega \frac{T_m}{4}} e^{j\omega t}
\end{aligned} \tag{2}
$$

Therefore, the total output signal mixed by the four-phase clock signal can be calculated as

$$
\begin{aligned}
v_2^-(t)_{total} &= e^{-j\omega \frac{T_m}{4}} e^{j\omega t} \\
&= e^{-j\omega \frac{T_m}{4}} v_1^+(t)
\end{aligned} \tag{3}
$$

Similar derivation can be performed for the reverse direction and the final result is as follows:

$$
\begin{aligned}
v_1^-(t)_{total} &= -e^{-j\omega \frac{T_m}{4}} e^{j\omega t} \\
&= -e^{-j\omega \frac{T_m}{4}} v_2^+(t)
\end{aligned} \tag{4}
$$

Equations (3) and (4) show that the ideal gyrator using a 25% duty-cycle clock signal is lossless and appears externally as linear time invariant (LTI), which has the same results compared with the gyrator using a 50% duty-cycle clock signal [34].

2.3. System-Level Analysis

In order to provide a more straightforward and accurate analysis of the gyrator's performance, a system-level analysis method is proposed in this section.

Starting from Equation (1), when the LO signal is mixed with an input signal expressed as $v_1^+(t) = e^{j\omega t}$, the output signal $h(t)$ is shown as follows:

$$
\begin{aligned}
h(t) &= LO(t) \times v_1^+(t) \\
&= \tfrac{2}{\pi} \sum_{n=1}^{\infty} \frac{\sin \frac{(2n-1)\pi}{4}}{2n-1} \left(e^{j(\omega+(2n-1)\omega_m)t} + e^{j(\omega-(2n-1)\omega_m)t} \right)
\end{aligned} \tag{5}
$$

Now, if the output signal in Equation (5) is directly mixed again with the square wave in Equation (1), the output signal can be derived as

$$
\begin{aligned}
v_2^-(t) &= LO(t) \times v_1^+(t) \times LO(t) \\
&= \tfrac{2}{\pi} \sum_{n=1}^{\infty} \frac{\sin \frac{(2n-1)\pi}{4}}{2n-1} \left(e^{j(\omega+(2n-1)\omega_m)t} + e^{j(\omega-(2n-1)\omega_m)t} \right) \\
&\times \tfrac{2}{\pi} \sum_{n=1}^{\infty} \frac{\sin \frac{(2n-1)\pi}{4}}{2n-1} \left(e^{j(2n-1)\omega_m t} + e^{-j(2n-1)\omega_m t} \right) \\
&= \tfrac{1}{2} e^{j\omega t}
\end{aligned} \tag{6}
$$

Then, the total output signal mixed by the four-phase clock signal can be calculated as

$$
v_2^-(t)_{total} = v_1^+(t) \tag{7}
$$

From Equations (5)–(7), it is found that the system illustrated above is independent with mixing products and is externally LTI and reciprocal as

$$
v_2^-(t-\tau) = LO(t-\tau) \times v_1^+(t-\tau) \times LO(t-\tau) \tag{8}
$$

and

$$
\begin{aligned}
& a v_{2,a}^-(t-\tau) + b v_{2,b}^-(t-\tau) \\
&= LO(t-\tau) \times \left(a v_{1,a}^+(t-\tau) + b v_{1,b}^+(t-\tau) \right) \times LO(t-\tau)
\end{aligned} \tag{9}
$$

The system derived in Equations (5)–(9) can be illustrated as in Figure 2a,b. From another point of view, the system with time delay τ in Figure 2b can be separated into two sub-systems in the forward and reverse direction if the time delay is performed by the time delay module internally, as shown in Figure 3a,b. Note that the time delay module should be non-reciprocal and the system implements a time delay of $\pm\tau$ in the forward and reverse direction.

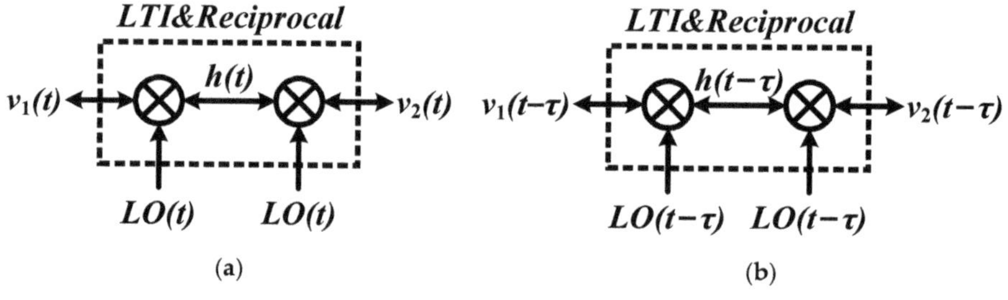

Figure 2. Doubly square-wave mixing system with LTI and reciprocal property: (**a**) original case; (**b**) with time delay τ.

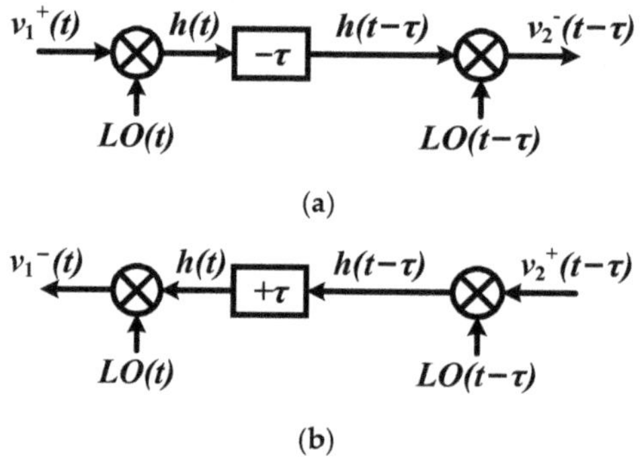

Figure 3. Doubly square-wave mixing system with time delay in the internal node: (**a**) forward direction; (**b**) reverse direction.

If the time delay module is implemented by a transmission line, it must be reciprocal in both directions. Therefore, in order to implement the system shown in Figure 3a,b using a transmission line, the relationship shown below is required:

$$h(t+\tau) = h(t-\tau) \tag{10}$$

Considering the frequency components in the internal node after mixing expressed as

$$\omega \pm (2n-1)\omega_m \tag{11}$$

when $\tau = T_m/n$ and the frequency of the input signal is the harmonics of ω_m, the ratio n that satisfies Equation (10) is analyzed. Under such a circumstance, the frequency components in the internal node are the harmonics of ω_m, and if the minimum frequency component meets the requirement of Equation (10), all the higher-order harmonics will be satisfied. Thus, the minimum frequency component is analyzed here.

When the frequency of the input signal is the even harmonics of ω_m, the minimum frequency component in the internal node is ω_m, i.e.,

$$\begin{gathered} 2\tau = T_{\mathrm{m}} \\ n = 2 \end{gathered} \tag{12}$$

In this case, a reciprocal phase shift of $T_m/2$ in the forward and reverse direction is implemented.

When the frequency of the input signal is the odd harmonics of ω_m, the minimum frequency component is $2\omega_m$, i.e.,

$$\begin{aligned} 2\tau &= \frac{T_m}{2} \\ n &= 4 \end{aligned} \tag{13}$$

In this case, a non-reciprocal phase shift of $\pm T_m/4$ in the forward and reverse direction is implemented and the function of a gyrator is realized, which is used in the designed circulator.

3. Implementation

A comprehensive research framework is shown in Figure 4. The implementation details are illustrated in this section and the design procedure of the artificial transmission line is followed according to [37].

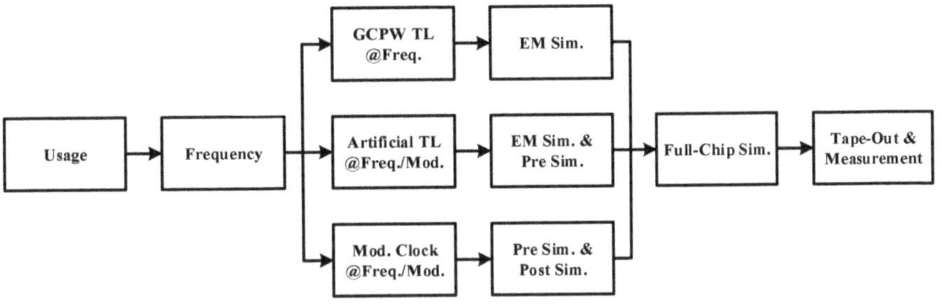

Figure 4. Comprehensive research framework of proposed circulator.

3.1. Generation of 25% Duty-Cycle Clock Signal

Conventionally, a 25% duty-cycle clock signal is generated using a divide-by-2 frequency divider and AND gates, as shown in Figure 5, which results in high power consumption and poor phase-noise performance. In order not to deteriorate the performance of the whole full-duplex communication transceiver and implement a high-frequency, high-performance 25% duty-cycle clock signal, the generator that outputs the 25% duty-cycle clock signal directly after the divide-by-2 divider is used in this design, as shown in Figure 6.

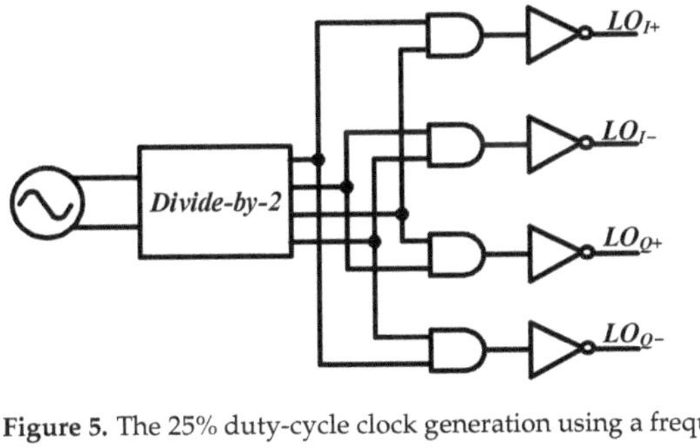

Figure 5. The 25% duty-cycle clock generation using a frequency divider and AND gates.

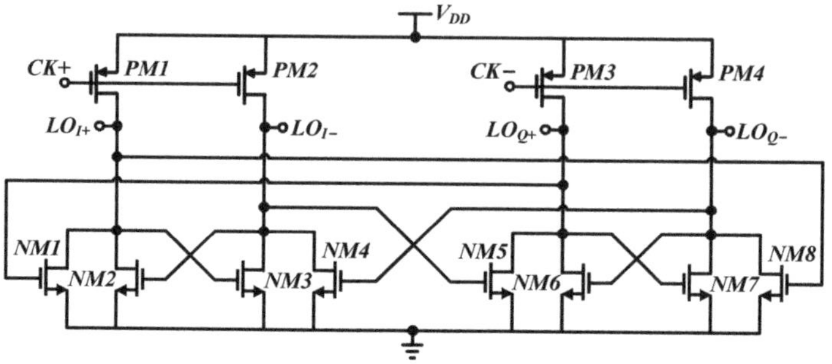

Figure 6. The 25% duty-cycle clock generation directly after the divide-by-2 frequency divider.

The sinusoidal input signal with a frequency of $2f_{LO}$ is injected using the ground-signal-ground (GSG) pad and the single-to-differential conversion is performed on the chip. A four-stage inverter buffer with resistor feedback in the second and fourth stages is applied directly after the 25% duty-cycle quadrature output. In order to mitigate the influence of the long-distance layout routing of the clock signal to the switches, another two-stage inverter buffer with resistor feedback in the second stage is placed near to the switches in the layout design.

A $2f_{LO}$ (i.e., 8.4 GHz) sinusoidal signal is used as the input of the generator. The clock signal generator consumes power of 47.3 mW, which includes 3.9 mW from the generator core and 43.4 mW from all the buffers. After the inverter buffer chain, a four-phase 25% duty-cycle square-wave clock signal, of which the rising time is approximately 1/12 of the modulation period, is generated.

3.2. Proposed Circulator

The passive on-chip non-reciprocal CMOS circulator is proposed in this section, as shown in Figure 7, embedded with the gyrator using a 25% duty-cycle clock signal. The operating frequency is designed to be 12.6 GHz with a 4.2 GHz modulation clock. In order to absorb the parasitic capacitance of the switches to the $\lambda/8$ transmission line at 12.6 GHz, a symmetric structure is utilized in this design.

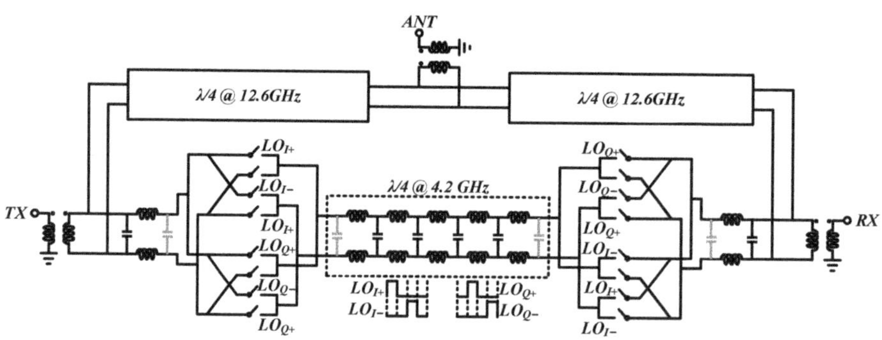

Figure 7. Proposed circulator using 25% duty-cycle clock signal.

Considering the fabrication technology limitations, bulk CMOS transistors without a thick gate are used as the mixer switches, of which the parasitic capacitance is absorbed in the gyrator symmetrically.

The $\lambda/8$ transmission line at 12.6 GHz between the TX/RX port and gyrator is implemented by the *LC* lumped element, and the $\lambda/4$ transmission line at 12.6 GHz between the TX/RX port and ANT port is implemented by a grounded coplanar waveguide (GCPW). Taking the size, insertion loss, and bandwidth into consideration, the $\lambda/4$ transmission line at 4.2 GHz in the gyrator is realized by a 5-stage *LC* π-type lumped element transmission line. The simulated Bragg frequency is approximately 42 GHz, which can tolerate up to seven-times mixing products and is limited by the operating frequency and process used. Recently, the *LC* π-type lumped element transmission line has been modified to obtain a lower insertion loss, such as a hybrid π-lattice structure [42], band-pass filter structure [43], all-pass filter structure [39], etc., which are also suitable in the proposed circulator. In addition, in order to enable single-ended measurements, baluns are attached to the proposed design.

4. Measurements

The proposed on-chip passive circulator is fabricated in a 90-nm bulk CMOS technology. The die micro-photograph of the chip is shown in Figure 8, and the total chip area is 1.55 mm × 1.15 mm with the TX and RX baluns and bonding pads excluded (ANT balun is difficult to de-embed, as shown in Figure 7). The measurement is performed by an RF probe with the die mounted on a PCB board. The power supply and ground are wire bonded to the board.

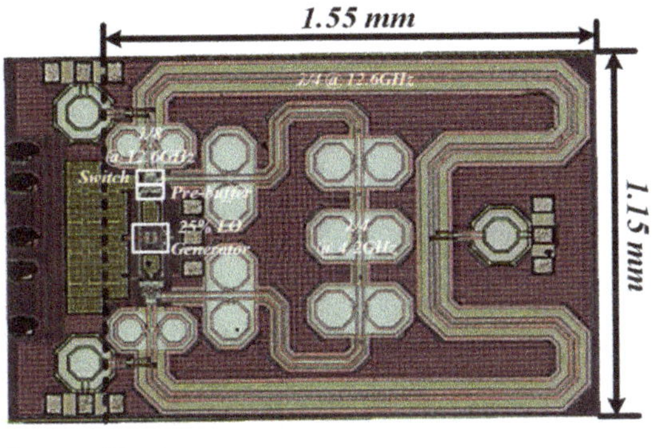

Figure 8. Die photo.

Four GSG pads are included on the chip for the probing test. Three of them are TX, ANT, and RX terminals and the other one is for clock signal input. The 8.4 GHz sinusoidal input clock signal is generated using a millimeter-wave signal generator and the single-to-differential conversion is implemented on the chip, as illustrated above. S-parameters are measured using a vector network analyzer with two ports probed at a time and the third port terminated with 50 Ω. On-chip calibration of the vector network analyzer is performed ahead of S-parameter measurement. The measured S-parameters of TX-to-ANT (port 1 and port 2), ANT-to-RX (port 2 and port 3) and TX-to-RX (port 1 and port 3) are shown in Figure 9a–c. The insertion loss (IL) is 3.9 dB from TX to ANT (S_{21}) and 4.0 dB from ANT to RX (S_{32}). The TX-to-RX isolation (S_{31}) is >17.2 dB. The S-parameter results are concluded without any on-chip reconfiguration and ANT port impedance tuning, which leads to the reflection at the ANT port and influences the isolation performance. The bandwidth of 1-dB IL degradation is 2.7 GHz (21.4%), corresponding to an isolation of 17.2 to 21.7 dB.

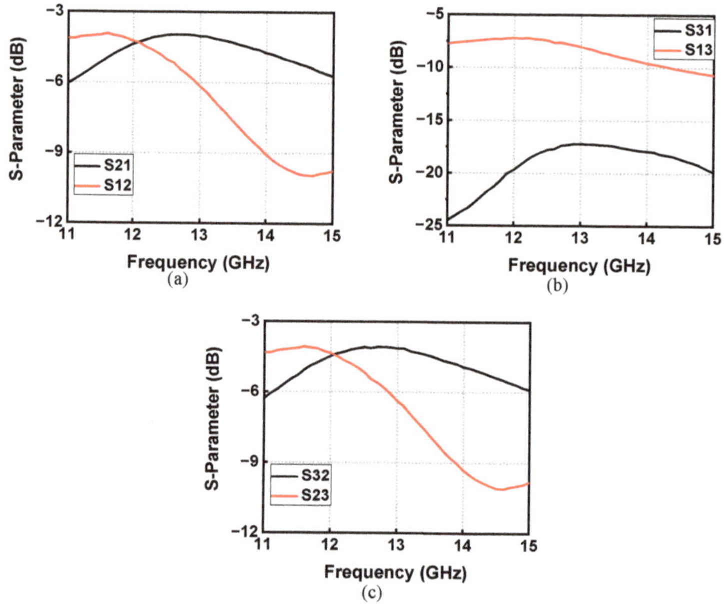

Figure 9. S-parameter measurements: (**a**) S_{21} and S_{12}; (**b**) S_{31} and S_{13}; (**c**) S_{32} and S_{23}.

The linearity of the proposed circulator is measured with a two-tone (12.6 GHz and 12.7 GHz) third-order intercept point (IP3) test. The fundamental output amplitude is extrapolated along a line of unity slope and the third-order intermodulation product (IM3) is extrapolated along a line with a slope of three, with the results exhibited in Figure 10a,b. It shows that an input third-order intercept point (IIP3) of 19.7 dBm from TX to ANT and 20.0 dBm from ANT to RX are achieved.

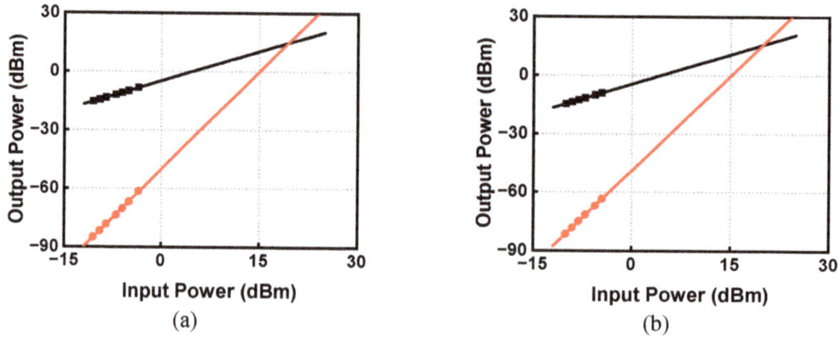

Figure 10. IIP3 measurements (**a**) TX to ANT; (**b**) ANT to RX.

A performance summary and comparison is shown in Table 4. Compared to the prior work, this paper has implemented an on-chip non-reciprocal passive circulator using bulk CMOS technology operating at the Ku band. Thanks to the high-speed 25% duty-cycle clock signal generation circuit, the proposed circulator realizes competitive performance in isolation and linearity compared with the designs using SOI technology, which makes the cost lower. Although process limitations exist, the insertion loss from TX to ANT and from ANT to RX only deteriorates slightly compared to the design in CMOS SOI, which is significantly influenced by the process and operating frequency.

Table 4. Comparison table with the state-of-the-art.

	This Work	[37]	[42]	[43]	[47]	[23]
Type	Passive Circulator	Passive Circulator	Passive Circulator	Passive Circulator	Passive Circulator	Active Quasi-Circulator
Technology	90-nm CMOS	45-nm CMOS SOI	180-nm CMOS SOI	45-nm CMOS SOI	180-nm CMOS	180-nm CMOS
Frequency (GHz) *	11.6–14.3	22.7–27.3	0.86–1.08	50–56.8	0.89–0.92	1–7 ****
Designed Center Frequency (GHz)	12.6	25	0.95	54	0.91	N/A
Reconfiguration /Tuning	No	Clock Phase Tuning	Antenna Tuning	Clock Phase Tuning	External Reconfigurable Impedance Tuning	No
TX-ANT Transmission (dB)	−3.9 @12.6 GHz	−3.3 @25 GHz	−2.1 @0.95 GHz	−3.6(1.2V V_{DD})/−3.9(1V V_{DD}) @54 GHz	−4.8 @0.91 GHz	>−10
ANT-RX Transmission (dB)	−4.0 @12.6 GHz	−3.2 @24.7 GHz	−2.9 @0.95 GHz	−3.1 @54 GHz	−4.8 @0.91 GHz	>−9
TX-RX Isolation (dB) **	> 17.2	> 18.3	> 25	> 20	> 20	>36
Isolation Bandwidth ***	21.4%	18.4%	17%	14.6%	3.4%	N/A
TX-ANT IIP3 (dBm)	19.7	20.1	50.025	19.43	6.1	9.7
ANT-RX IIP3 (dBm)	20	19.9	36.9	19.03	6	3.5
Area (mm^2)	1.55 × 1.15	1.8 × 1.2	4.6 × 3.6	1.57 × 1.1	36	1.03 × 0.55

* Frequency range of 1-dB IL degradation; ** Isolation in the frequency range; *** Frequency range divided by designed center frequency; **** Secured isolation bandwidth.

5. Limitations of the Study and Recommendations

Due to the lack of a qualified pre-amplifier, the followings are absent: noise figure, P1dB, and spurious tone measurement. However, P1dB and spurious tones can be reflected by IIP3 to some extent, and the noise figure should be slightly larger than the insertion loss in this passive circulator due to the phase noise of the clock signal.

From the current measurement results, it is recommended to use this circulator in a full-duplex system operating at around 10–20 GHz and requiring a low cost. Although the insertion loss is slightly larger than designs using CMOS SOI technology, other measurement results are competitive, and, moreover, the isolation bandwidth is slightly larger without any tuning method. The proposed design is a promising candidate for Ku-band full-duplex satellite communication.

6. Conclusions

This paper presents an on-chip non-reciprocal passive circulator using bulk CMOS technology with spatio-temporal conductivity modulation operating at the Ku band. Fabricated in a 90-nm bulk CMOS technology, the proposed 12.6-GHz circulator implements an insertion loss of 3.9 dB from TX to ANT and of 4.0 dB from ANT to RX, and with 1-dB degradation over 2.7 GHz, which corresponds to a bandwidth of 21.4%. A > 17.2 dB isolation is achieved, which ranges from 17.2 dB to 21.7 dB in the 1-dB degradation insertion loss bandwidth. Linearity performance is also competitive compared to the prior work using SOI technology without antenna impedance tuning. The proposed circulator consumes an area of 1.55 mm × 1.15 mm and is suitable for low-cost RFIC design.

Author Contributions: Conceptualization, J.G. and J.W.; methodology, X.W. and F.H.; software, J.G.; validation, J.G., W.G. and F.H.; formal analysis, W.G.; investigation, X.W. and J.W.; resources, J.G. and X.W.; data curation, F.H.; writing—original draft preparation, J.G.; writing—review and editing, X.W. and W.G.; visualization, W.G.; supervision, J.W.; project administration, J.W.; funding acquisition, X.W. and J.W. All authors have read and agreed to the published version of the manuscript.

Funding: This work was funded in part by the National Natural Science Foundation of China under Grant 62101038, in part by the Beijing Nova Program of Science and Technology under Grant Z191100001119078, and in part by the China Postdoctoral Science Foundation under Grant 2021M700406.

Data Availability Statement: Not applicable.

Conflicts of Interest: The authors declare no conflict of interest.

References

1. Debaillie, B.; van Liempd, B.; Hershberg, B.; Craninckx, J.; Rikkinen, K.; van den Broek, D.J.; Klumperink, E.A.M.; Nauta, B. In-band Full-Duplex Transceiver Technology for 5G Mobile Networks. In Proceedings of the ESSCIRC Conference 2015—41st European Solid-State Circuits Conference (ESSCIRC), Graz, Austria, 14–18 September 2015.
2. Giordani, M.; Polese, M.; Mezzavilla, M.; Rangan, S.; Zorzi, M. Toward 6G Networks: Use Cases and Technologies. *IEEE Commun. Mag.* **2020**, *58*, 55–61. [CrossRef]
3. Reiskarimian, N.; Dinc, T.; Zhou, J.; Chen, T.; Dastjerdi, M.B.; Diakonikolas, J.; Zussman, G.; Krishnaswamy, H. One-way ramp to a two-way highway: Integrated magnetic-free nonreciprocal antenna interfaces for full-duplex wireless. *IEEE Microw. Mag.* **2019**, *20*, 56–75. [CrossRef]
4. Zhou, J.; Reiskarimian, N.; Diakonikolas, J.; Dinc, T.; Chen, T.; Zussman, G.; Krishnaswamy, H. Integrated full duplex radios. *IEEE Commun. Mag.* **2017**, *55*, 142–151. [CrossRef]
5. Barneto, C.B.; Riihonen, T.; Turunen, M.; Anttila, L.; Fleischer, M.; Stadius, K.; Ryynanen, J.; Valkama, M. Full-Duplex OFDM Radar With LTE and 5G NR Waveforms: Challenges, Solutions, and Measurements. *IEEE Trans. Microw. Theory Tech.* **2019**, *67*, 4042–4054. [CrossRef]
6. Biedka, M.; Wang, Y.E.; Xu, Q.M.; Li, Y. Full-Duplex RF Front Ends: From Antennas and Circulators to Leakage Cancellation. *IEEE Microw. Mag.* **2019**, *20*, 44–55. [CrossRef]
7. Dinc, T.; Chakrabarti, A.; Krishnaswamy, H. A 60 GHz CMOS Full-Duplex Transceiver and Link with Polarization-Based Antenna and RF Cancellation. *IEEE J. Solid-State Circuits* **2016**, *51*, 1125–1140. [CrossRef]
8. Chi, T.; Park, J.S.; Li, S.; Wang, H. A 64GHz Full-Duplex Transceiver Front-End with an On-Chip Multifeed Self-Interference-Canceling Antenna and an All-Passive Canceler Supporting 4Gb/s Modulation in One Antenna Footprint. In Proceedings of the 2018 IEEE International Solid—State Circuits Conference—(ISSCC), San Francisco, CA, USA, 11–15 February 2018.
9. Le, A.T.; Tran, L.C.; Huang, X.; Guo, Y.J. Analog Least Mean Square Loop With I/Q Imbalance for Self-Interference Cancellation in Full-Duplex Radios. *IEEE Trans. Veh. Technol.* **2019**, *68*, 9848–9860. [CrossRef]
10. Le, A.T.; Tran, L.C.; Huang, X.; Guo, Y.J. Beam-Based Analog Self-Interference Cancellation in Full-Duplex MIMO Systems. *IEEE Trans. Wirel. Commun.* **2020**, *19*, 2460–2471. [CrossRef]
11. Zhou, J.; Chuang, T.-H.; Dinc, T.; Krishnaswamy, H. Integrated Wideband Self-Interference Cancellation in the RF Domain for FDD and Full-Duplex Wireless. *IEEE J. Solid-State Circuits* **2015**, *50*, 3015–3031. [CrossRef]
12. Roberts, I.P.; Jain, H.B.; Vishwanath, S. Equipping Millimeter-Wave Full-Duplex with Analog Self-Interference Cancellation. In Proceedings of the 2020 IEEE International Conference on Communications Workshops (ICC Workshops), Dublin, Ireland, 7–11 June 2020.
13. Erdem, M.; Gurbuz, O.; Ozkan, H. A Residual Scheme for Digital Self-Interference Cancellation in Full Duplex Communication. In Proceedings of the 2020 IEEE International Conference on Communications Workshops (ICC Workshops), Dublin, Ireland, 7–11 June 2020.
14. Ershadi, A.; Entesari, K. A 0.5-to-3.5-GHz Full-Duplex Mixer-First Receiver With Cartesian Synthesized Self-Interference Suppression Interface in 65-nm CMOS. *IEEE Trans. Microw. Theory Tech.* **2020**, *68*, 1995–2010. [CrossRef]
15. Zhang, T.; Su, C.; Najafi, A.; Rudell, J.C. Wideband Dual-Injection Path Self-Interference Cancellation Architecture for Full-Duplex Transceivers. *IEEE J. Solid-State Circuits* **2018**, *53*, 1563–1576. [CrossRef]
16. Hamza, A.; Hill, C.; AlShammary, H.; Buckwalter, J. High-Rejection RF Code Domain Receivers for Simultaneous Transmit and Receive Applications. *IEEE J. Solid-State Circuits* **2020**, *55*, 1909–1921. [CrossRef]

17. Waheed, M.Z.; Korpi, D.; Anttila, L.; Kiayani, A.; Kosunen, M.; Stadius, K.; Campo, P.P.; Turunen, M.; Allen, M.; Ryynanen, J.; et al. Passive Intermodulation in Simultaneous Transmit–Receive Systems: Modeling and Digital Cancellation Methods. *IEEE Trans. Microw. Theory Tech.* **2020**, *68*, 3633–3652. [CrossRef]
18. Haq, F.U.; Englund, M.; Antonov, Y.; Stadius, K.; Kosunen, M.; Ryynänen, J.; Östman, K.B.; Koli, K. Full-Duplex Wireless Transceiver Self-Interference Cancellation Through FD-SOI Buried-Gate Signaling. In Proceedings of the 2018 IEEE International Symposium on Circuits and Systems (ISCAS), Florence, Italy, 27–30 May 2018.
19. Mikhemar, M.; Darabi, H.; Abidi, A. A tunable integrated duplexer with 50dB isolation in 40nm CMOS. In Proceedings of the 2009 IEEE International Solid-State Circuits Conference—Digest of Technical Papers, San Francisco, CA, USA, 8–12 February 2009.
20. Mikhemar, M.; Darabi, H.; Abidi, A.A. A Multiband RF Antenna Duplexer on CMOS: Design and Performance. *IEEE J. Solid-State Circuits* **2013**, *48*, 2067–2077. [CrossRef]
21. Tanaka, S.; Shimomura, N.; Ohtake, K. Active circulators—The realization of circulators using transistors. *Proc. IEEE* **1965**, *53*, 260–267. [CrossRef]
22. Mung, S.W.Y.; Chan, W.S. The Challenge of Active Circulators: Design and Optimization in Future Wireless Communication. *IEEE Microw. Mag.* **2019**, *20*, 55–66. [CrossRef]
23. Tang, B.; Gui, X.; Xu, J.; Xia, Q.; Geng, L. A Dual Interference-Canceling Active Quasi-Circulator Achieving 36-dB Isolation Over 6-GHz Bandwidth. *IEEE Microw. Wirel. Compon. Lett.* **2019**, *29*, 409–411. [CrossRef]
24. Wang, S.; Lee, C.-H.; Wu, Y.-B. Fully Integrated 10-GHz Active Circulator and Quasi-Circulator Using Bridged-T Networks in Standard CMOS. *IEEE Trans. Very Large Scale Integr. (VLSI) Syst.* **2016**, *24*, 3184–3192. [CrossRef]
25. Fang, K.; Buckwalter, J.F. A Tunable 5–7 GHz Distributed Active Quasi-Circulator With 18-dBm Output Power in CMOS SOI. *IEEE Microw. Wirel. Compon. Lett.* **2017**, *27*, 998–1000. [CrossRef]
26. Nagulu, A.; Reiskarimian, N.; Krishnaswamy, H. Non-reciprocal electronics based on temporal modulation. *Nat. Electron.* **2020**, *3*, 241–250. [CrossRef]
27. Kodera, T.; Sounas, D.L.; Caloz, C. Magnetless Nonreciprocal Metamaterial (MNM) Technology: Application to Microwave Components. *IEEE Trans. Microw. Theory Tech.* **2013**, *61*, 1030–1042. [CrossRef]
28. Fleury, R.; Sounas, D.L.; Sieck, C.F.; Haberman, M.R.; Alù, A. Sound isolation and giant linear nonreciprocity in a compact acoustic circulator. *Science* **2014**, *343*, 516–519. [CrossRef] [PubMed]
29. Estep, N.A.; Sounas, D.L.; Soric, J.; Alù, A. Magnetic-free non-reciprocity and isolation based on parametrically modulated coupled-resonator loops. *Nat. Phys.* **2014**, *10*, 923–927. [CrossRef]
30. Qin, S.; Xu, Q.; Wang, Y.E. Nonreciprocal Components With Distributedly Modulated Capacitors. *IEEE Trans. Microw. Theory Tech.* **2014**, *62*, 2260–2272. [CrossRef]
31. Mahmoud, A.M.; Davoyan, A.R.; Engheta, N. All-passive nonreciprocal metastructure. *Nat. Commun.* **2015**, *6*, 8359. [CrossRef]
32. Reiskarimian, N.; Krishnaswamy, H. Magnetic-free non-reciprocity based on staggered commutation. *Nat. Commun.* **2016**, *7*, 11217. [CrossRef]
33. Dinc, T.; Tymchenko, M.; Nagulu, A.; Sounas, D.; Alù, A.; Krishnaswamy, H. Synchronized conductivity modulation to realize broadband lossless magnetic-free non-reciprocity. *Nat. Commun* **2017**, *8*, 795. [CrossRef]
34. Nagulu, A.; Dinc, T.; Xiao, Z.; Tymchenko, M.; Sounas, D.L.; Alù, A.; Krishnaswamy, H. Nonreciprocal Components Based on Switched Transmission Lines. *IEEE Trans. Microw. Theory Tech.* **2018**, *66*, 4706–4725. [CrossRef]
35. Kord, A.; Sounas, D.L.; Alu, A. Microwave Nonreciprocity. *Proc. IEEE* **2020**, *108*, 1728–1758. [CrossRef]
36. Reiskarimian, N.; Zhou, J.; Krishnaswamy, H. A CMOS Passive LPTV Nonmagnetic Circulator and Its Application in a Full-Duplex Receiver. *IEEE J. Solid-State Circuits* **2017**, *52*, 1358–1372. [CrossRef]
37. Dinc, T.; Nagulu, A.; Krishnaswamy, H. A Millimeter-Wave Non-Magnetic Passive SOI CMOS Circulator Based on Spatio-Temporal Conductivity Modulation. *IEEE J. Solid-State Circuits* **2017**, *52*, 3276–3292. [CrossRef]
38. Ruffino, A.; Peng, Y.; Sebastiano, F.; Babaie, M.; Charbon, E. A 6.5-GHz Cryogenic All-Pass Filter Circulator in 40-nm CMOS for Quantum Computing Applications. In Proceedings of the 2019 IEEE Radio Frequency Integrated Circuits Symposium (RFIC), Boston, MA, USA, 2–4 June 2019.
39. Ruffino, A.; Peng, Y.; Sebastiano, F.; Babaie, M.; Charbon, E. A Wideband Low-Power Cryogenic CMOS Circulator for Quantum Applications. *IEEE J. Solid-State Circuits* **2020**, *55*, 1224–1238. [CrossRef]
40. Reiskarimian, N.; Khorshidian, M.; Krishnaswamy, H. Inductorless, Widely Tunable N-Path Shekel Circulators Based on Harmonic Engineering. *IEEE J. Solid-State Circuits* **2021**, *56*, 1425–1437. [CrossRef]
41. Nagulu, A.; Mekkawy, A.; Tymchenko, M.; Sounas, D.; Alù, A.; Krishnaswamy, H. Ultra-Wideband Switched-Capacitor Delays and Circulators—Theory and Implementation. *IEEE J. Solid-State Circuits* **2021**, *56*, 1412–1424. [CrossRef]
42. Nagulu, A.; Krishnaswamy, H. Non-Magnetic CMOS Switched-Transmission-Line Circulators With High Power Handling and Antenna Balancing: Theory and Implementation. *IEEE J. Solid-State Circuits* **2019**, *54*, 1288–1303. [CrossRef]
43. Nagulu, A.; Krishnaswamy, H. 28.5 Non-Magnetic 60GHz SOI CMOS Circulator Based on Loss/Dispersion-Engineered Switched Bandpass Filters. In Proceedings of the 2019 IEEE International Solid- State Circuits Conference—(ISSCC), San Francisco, CA, USA, 17–21 February 2019.
44. Nagulu, A.; Chen, T.; Zussman, G.; Krishnaswamy, H. Multi-Watt, 1-GHz CMOS Circulator Based on Switched-Capacitor Clock Boosting. *IEEE J. Solid-State Circuits* **2020**, *55*, 3308–3321. [CrossRef]

45. Min, B.; Hwang, J. Fully Integrated Ultra-Wideband Differential Circulator Based on Sequentially Switched Delay Line in 28-nm FDSOI CMOS. In Proceedings of the 2022 IEEE Radio Frequency Integrated Circuits Symposium—(RFIC), Denver, CO, USA, 19–21 June 2022.
46. Tellegen, B.D.H. The gyrator, a new electric network element. *Philips Res. Rep.* **1948**, *3*, 81–101.
47. Kord, A.; Tymchenko, M.; Sounas, D.L.; Krishnaswamy, H.; Alu, A. CMOS Integrated Magnetless Circulators Based on Spatiotemporal Modulation Angular-Momentum Biasing. *IEEE Trans. Microw. Theory Tech.* **2019**, *67*, 2649–2662. [CrossRef]

micromachines

Article

A D-Band Direct-Conversion IQ Receiver with 28 dB CG and 7.3 dB NF in 130 nm SiGe Process

Fei He , Yuhan Ding, Zhongchen Xu, Menghu Ni, Yibo Tian, Zhenyi Zhang , Zhixiang Shi, Kailei Wang, Qian Xie and Zheng Wang *

School of Electronic Science and Engineering (National Exemplary School of Microelectronics), University of Electronic Science and Technology of China, Chengdu 611731, China
* Correspondence: wangzheng@uestc.edu.cn

Abstract: In this paper, a D-band direct conversion IQ receiver with on-chip multiplier chain is presented. The D-band LNA with gain-boosting and stagger-tunning technique is implemented to provide high gain and large bandwidth. X9 multiplier chain including Marchand balun and quadrature (90°) hybrid is employed to provide four path LO signal to drive IQ mixer. This receiver is implemented in a 130nm SiGe process and consumes a core area of 1.04 mm^2. From the experimental results, the proposed receiver exhibits a 20 GHz bandwidth from 150 GHz to 170 GHz, with CG of 28 dB and NF of 7.3 dB at 158 GHz.

Keywords: D-band receiver; direct conversion; on-chip multiplier chain; SiGe process; sub-THz

Citation: He, F.; Ding, Y.; Xu, Z.; Ni, M.; Tian, Y.; Zhang, Z.; Shi, Z.; Wang, K.; Xie, Q.; Wang, Z. A D-Band Direct-Conversion IQ Receiver with 28 dB CG and 7.3 dB NF in 130 nm SiGe Process. *Micromachines* **2023**, *14*, 87. https://doi.org/10.3390/mi14010087

Academic Editor: Zhiming Chen

Received: 8 December 2022
Revised: 23 December 2022
Accepted: 24 December 2022
Published: 29 December 2022

1. Introduction

Due to the abundant spectrum resources, THz and sub-THz systems attract more and more attention to high-speed communication systems and high-resolution radar [1–9]. Especially, with the emergence of advanced silicon processes which offer f_{max} above 400 GHz [10,11], the development of D-band (110–170GHz) front-ends grows rapidly. Nowadays, several D-band transceivers are being reported in communication and radar systems [12–15].

For the D-band receiver, high gain, low noise, and large bandwidth are the key requirements to strict the building block of the front-ends. To meet these above-mentioned requirements, in this paper, we present a D-band direct conversion IQ receiver with on-chip LO multiplier chain. Benefiting from the on-chip LO chain, this receiver has great potential to be integrated into the digital backend for communication and radar systems.

2. Receiver Architecture and Building Block

2.1. The Direct Conversion IQ Receiver

The architecture of the proposed D-band direct conversion IQ receiver is shown in Figure 1. It is implemented in a 130 nm SiGe process with an f_t/f_{max} of 300/450 GHz. This receiver integrated the LNA, LO chain, and IQ mixer. Due to the high gain and large bandwidth of the LNA, the D-band receiver can suppress the noise contribution due to all the blocks after the front-end LNA. To drive the IQ mixer, x9 LO multiplier chain with buffer amplifier, Marchand balun, and quadrature (90°) hybrid are implemented which can provide four paths of D-band signal.

2.2. The D-Band LNA

The schematic of the proposed LNA is shown in Figure 2. The design of the LNA mainly focuses on optimizing NF, bandwidth, and gain. Firstly, to obtain a sufficient power gain at such high frequency, four stage amplifier is cascaded in which the first and second stages are the cascade structure, with the third and the fourth stages being the common source structure. The sufficient power gain of the LNA is also beneficial to make the contribution of other blocks to the overall chain noise negligible. Secondly, for the NF optimization, since the first stage amplifier dominates the NF of the whole LNA, the source degeneration inductor TL_S and gain-boosted inductor TL_g are employed in the first stage amplifier, respectively, to obtain a low NF and high gain. Finally, to obtain a wide bandwidth, a gain control stagger-tuning technique is employed in this four-stage amplifier [16–19]. The peak gain of the first, second, third, and fourth-stage amplifiers is located at 160 GHz, 175 GHz, 145 GHz, and 145 GHz, respectively.

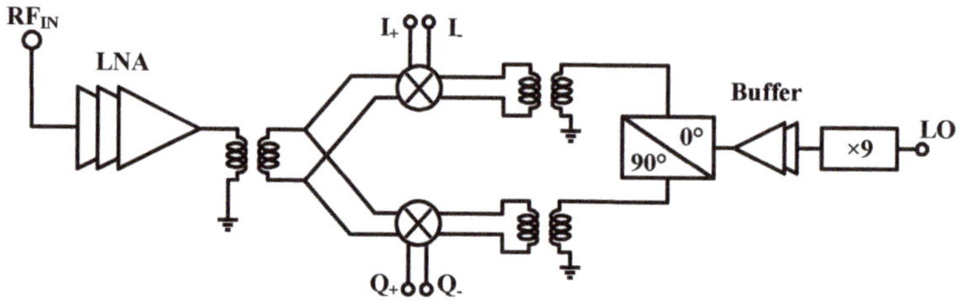

Figure 1. The D-band receiver structure.

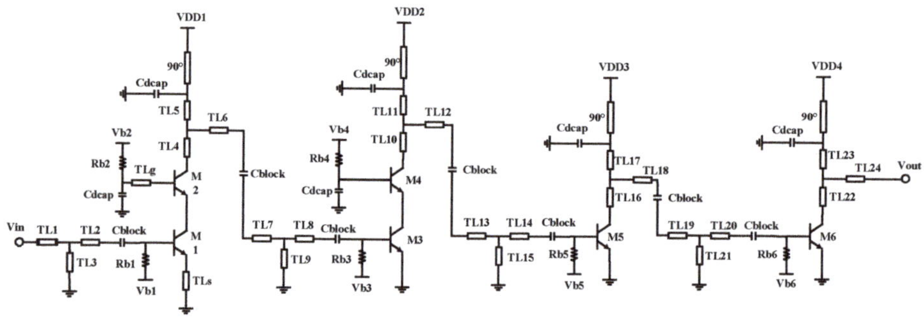

Figure 2. The schematic of the proposed wideband LNA.

The S-parameters and NF simulation results of the proposed wideband LNA are shown in Figure 3. From 150–170 GHz, the proposed LNA exhibits a flat gain range from 23.1 dB to 24.6 dB, it also achieved well input and output conjugate matching. Figure 3b shows the NF simulation result, from 150–170 GHz the proposed LNA exhibits a low NF value range from 6.93 dB to 7.45 dB.

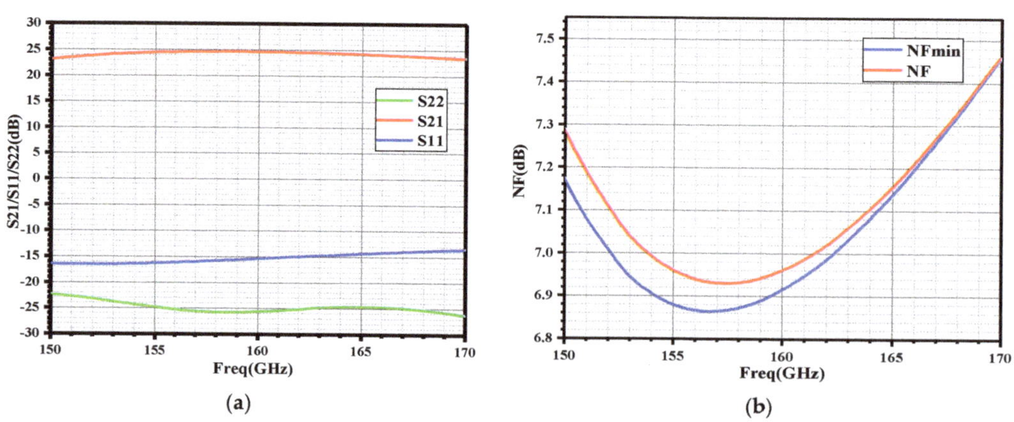

Figure 3. The simulation results of the four-stage LNA: (**a**) S-parameters; (**b**) NF.

2.3. The Mixer

The schematic of the D-band quadrature mixer is shown in Figure 4; it is based on a pseudo-differential Gilbert cell.

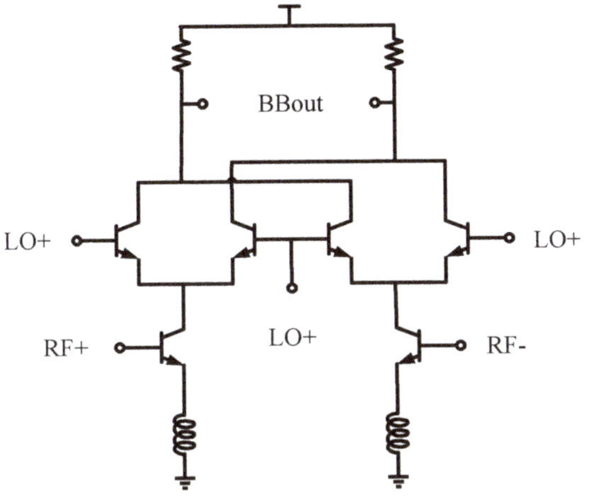

Figure 4. The schematic of the down-conversion mixer.

The pseudo-differential transconductance stage eliminates the linearity penalty caused by the tail-current source. As a down conversation mixer, the inductive emitter degeneration technique is employed at high frequency for enhancing linearity. The size of the transconductance stage was determined as minimum as possible to provide a high input impedance to relieve the stress of impedance matching. The size of the switching quad was also determined as a minimum to relax the LO requirement [20]. The mixer drives a 500 Ω load, a relatively large load that provides high voltage gain. Simulating the conversion gain, the S-parameter quadrature mixer results are shown in Figure 5.

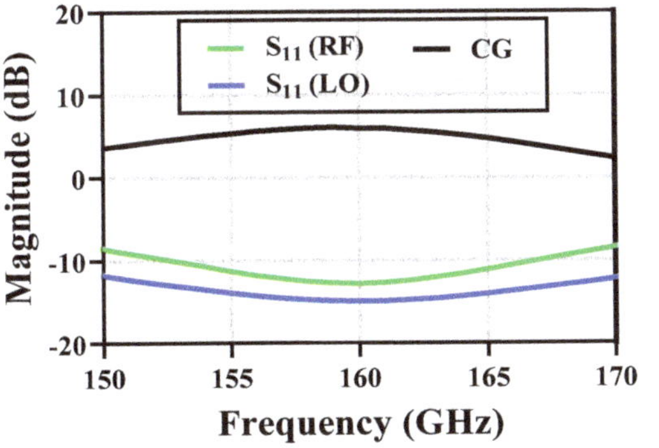

Figure 5. Simulation results of quadrature mixer.

2.4. The LO Multiplier Chain

The receiver employs the frequency multiplier chain to generate the 160 GHz LO signal with an external reference of 17 to 19 GHz. The schematic of the multiplier chain with buffer amplifiers is shown in Figure 6.

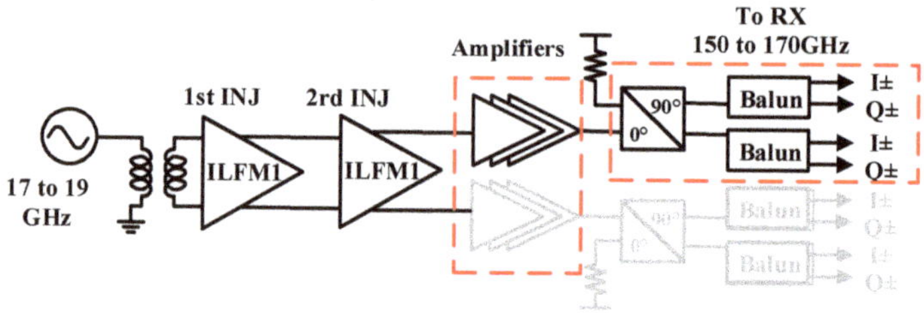

Figure 6. Schematic of the proposed LO chain.

The core part of the LO chain is a two-stage x3 injection-locked frequency multiplier (ILFM) to produce a x9 signal. Then, three-stage cascade amplifiers, 90° hybrid coupler, and Marchand balun are utilized to generate differential I/Q signals to drive mixers.

For the ILFM, Figure 7 represents the schematic of the transformer-based VCO. The x3 ILFM adopts the third-harmonic enhancement technique [21] in transformer-based LC VCO, ensuring the multiplier is working at the fundamental and third harmonic wave. Due to the nature that the frequency is locked at fundamental, with the amplifying and detecting function of the resonant tank in the collector, it is easier for the multiplier to be locked with a wider locking range. As a trade-off between large third harmonic and phase noise, Km is about 0.5 and 0.6 for the transformers of the first and second stages, which use the planar and overlay structure, respectively, to keep a relatively high and constant K_m and avoid the inductor self-resonance. The equivalent Q-factor of the transformer in Figure 7 can be calculated as:

$$Q_{eq} = \frac{1 + \alpha_p \alpha_s k_m^2 - \alpha_p \alpha_s}{\frac{\alpha_p}{Q_p} + \frac{\alpha_s}{Q_s} - \alpha_p \alpha_s \left(\frac{1}{Q_p} + \frac{1}{Q_s} \right)}, \tag{1}$$

where $\alpha_p = \omega^2 L_p C_p$, $\alpha_s = \omega^2 L_s C_s$, Q_p, and Q_s are the Q-factor for the primary and secondary winding. When equals to the fundamental and third harmonic, assuming $Q_p = Q_s$, (1) can be simplified to:

$$Q_{eq}\Big|_{\omega=\omega_{fund,3rd}} = Q_p \left(1 + \frac{2\alpha_s k_m^2}{1 + X - 2\alpha_s} \right), \tag{2}$$

where $X = L_s C_s / L_p C_p$. For a transformer based dual tank resonator, when $\omega = \omega_{fund}$, $\alpha_s < 1$ while $\omega = \omega_{3rd}$, $\alpha > 1/\sqrt{1 - k_m^2}$. From (2), it can be concluded that a smaller K_m brings a larger Q-factor and impendence at the third harmonic, which is desired for the enhancement and multiplying. However, a large K_m is still required for a high Q-factor at the fundamental, because the reduction in K_m makes the phase noise performance degrade.

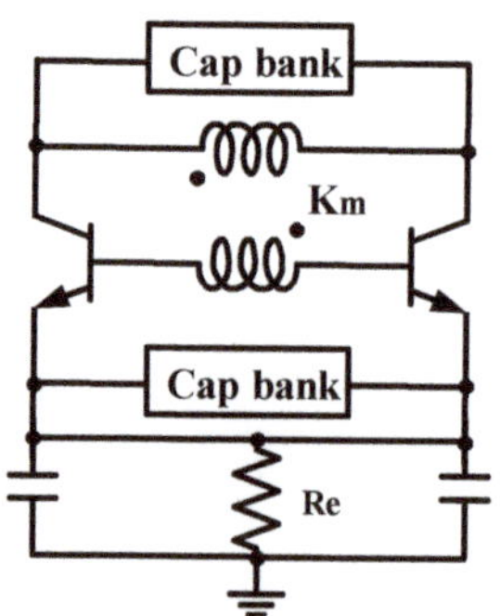

Figure 7. Schematic of the transformer-based VCO with third-harmonic extraction technique.

For the core part, the output power at the second amplifier reaches 0.3 dBm at 160 GHz with an input power of 6 dBm. The amplitude and phase mismatch at 150 to 170 GHz is <0.5 dB and <4.5°, respectively, with a –3 dBm input at the third amplifier. The total power consumption is 355 mW, in which the core consumes 91 mW from a 1.6 V supply, while the amplifier chain occupies 264 mW from a 3 V supply.

2.5. The Quadrature (90°) Hybrid

In the LO chain, to generate the quadrature-hybrid signal, a compact quadrature (90°) hybrid is implemented. The layout of the quadrature-hybrid is shown in Figure 8a. Co-planner waveguides (CPW) are chosen to realize the 50 Ω and 35 Ω transmission lines for their better shielding and higher integration at D-band. The 35 Ω CPWs are meandered to reduce the total size of the quadrature-hybrid. Simulation results show that the insertion loss is 1–1.5 dB and the phase/gain error is within ±3° and ±1 dB over 150–170 GHz.

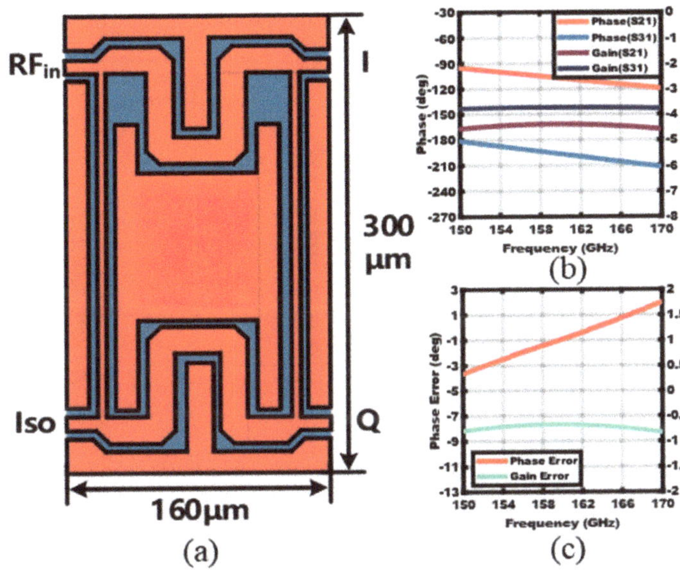

Figure 8. (**a**) Layout of the quadrature–hybrid, (**b**) simulated phase/gain response, and (**c**) simulated phase/gain error.

2.6. The Marchand Balun

Marchand balun is employed to achieve the function of single to differential signal conversion. The 3D structure picture and model picture of Marchand balun are shown in Figure 9.

The Marchand balun consists of two couplers. TopMetal2 and TopMetal1 are utilized to generate the coupler and the length of the transmission line in the coupler is 187 μm. The odd and even impedances of the quarter-wave-length coupler in the balun should be 26 and 96 Ω, respectively [22]. To save the chip area, this balun has been folded and it exhibits a symmetrical layout.

The proposed balun structure is EM-simulated by a fully-wave EM simulator. Figure 10a shows the simulation results of S-parameters. The S11 of balun is lower than −20 dB and the amplitude imbalance is less than 0.4 dB from 150 GHz to 170 GHz. Figure 10b shows the phase imbalance of the proposed balun; it exhibits a 6° phase imbalance at 160 GHz.

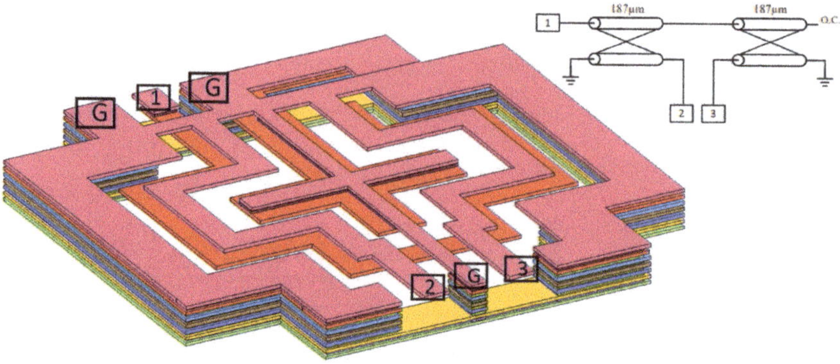

Figure 9. The 3D structure of the proposed Marchand balun.

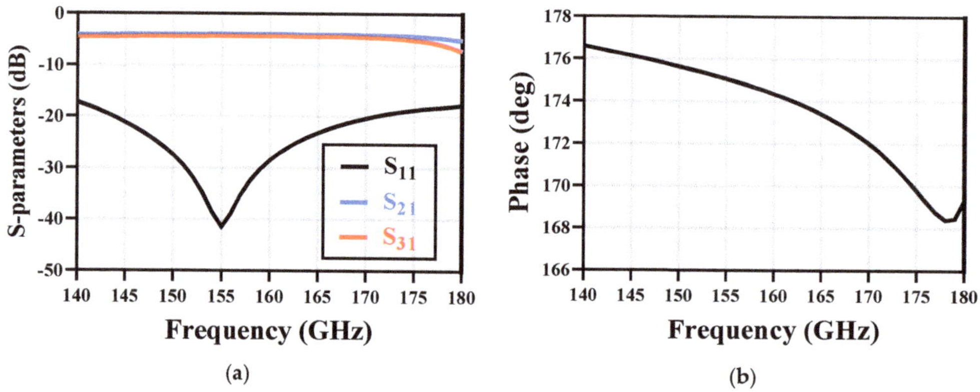

Figure 10. The EM simulation results of the Marchand balun: (**a**) S−parameters; (**b**) phase imbalance.

3. Experimental Result

The die photograph of the proposed D-band direct conversion IQ receiver is shown in Figure 11; it consumes a DC power of 428 mW and a total area of 1.04 mm^2 excluding all these RF and DC pads. Due to the measurement limitation, experimental results of conversion gain and linearity of the receiver are presented, with the results of the LNA part being measured and the results of the mixer part simulated.

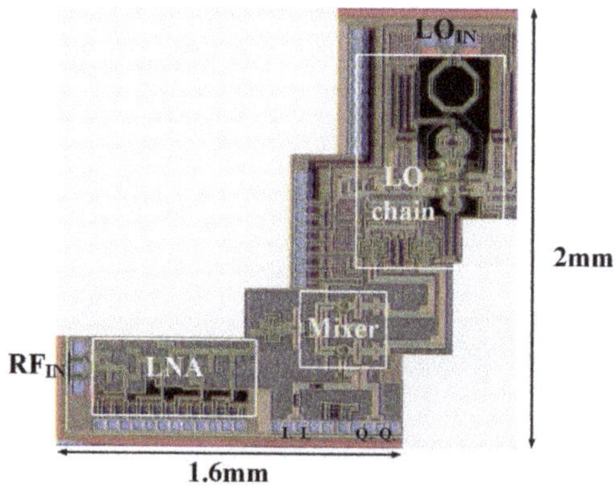

Figure 11. Die photograph of the D−band receiver chip.

3.1. The Conversion Gain

Conversion gain results at 1 GHz IF for the proposed receiver are shown in Figure 12. From 150 GHz to 170 GHz, the proposed receiver achieves an overall CG above 25 dB and a maximum CG of 28 dB was achieved at 158 GHz.

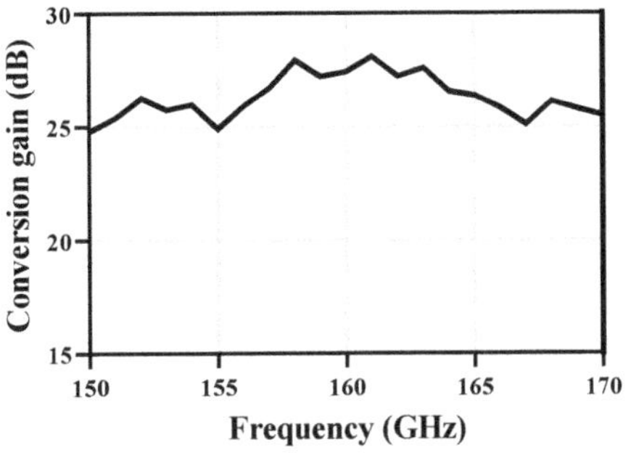

Figure 12. CG results at f_{IF} = 1 GHz.

3.2. NF

Due to the low noise and high gain performance of the D-band LNA in the receiver front-end, the proposed receiver exhibits a simulated NF result below 8 dB from 150 GHz to 170 GHz as shown in Figure 13. The lowest NF of 7.3 dB is achieved at 158 GHz.

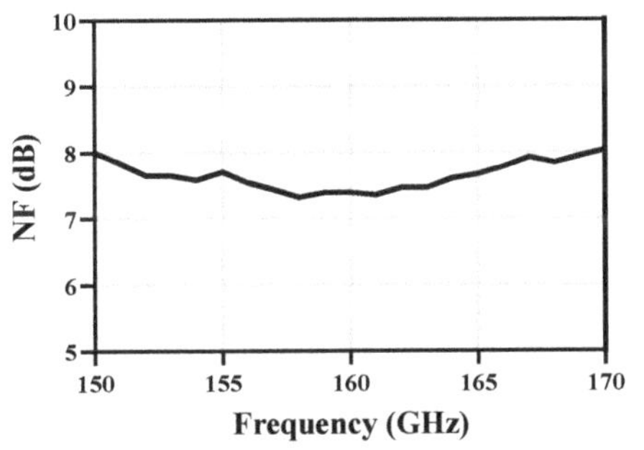

Figure 13. NF results at f_{IF} = 1GHz.

3.3. 1-dB Compression Point

1-dB compression point results at 160 GHz are shown in Figure 14. The proposed receiver achieved an input compression point of −19 dBm.

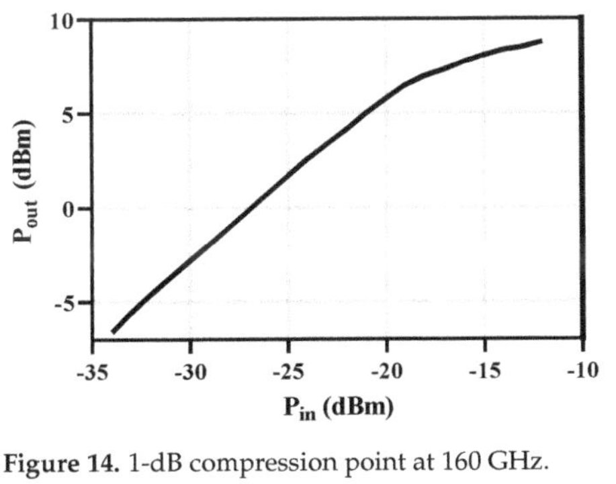

Figure 14. 1-dB compression point at 160 GHz.

3.4. S-Parameters

Measured and simulated S-parameters for the RF input of the proposed receiver are shown in Figure 15. It shows great matching on the RF input. The S_{11} is below −10 dB from 150 GHz to 170 GHz.

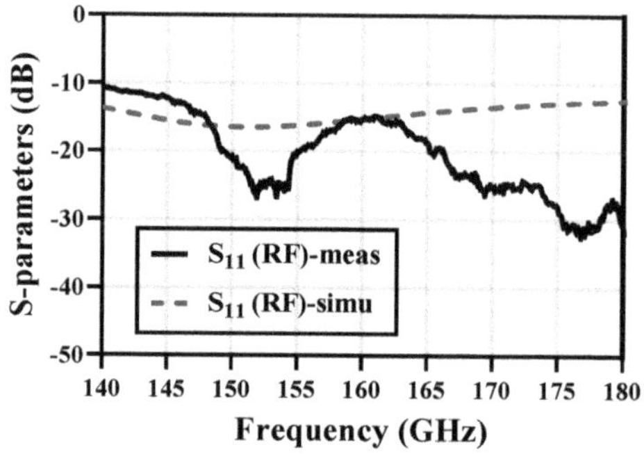

Figure 15. Simulated and measured S-parameters of the RF input.

4. Conclusions

This work presents a D-band direct conversion IQ receiver in a 130 nm SiGe process. The proposed fundamental receiver features a low NF D-band LNA and on-chip LO chain. The performance comparison between our work and the prior silicon-based D-band receiver is presented in Table 1. The proposed receiver is working at the high side of the D-band frequency range with large bandwidth, high gain, and low noise properties. It has great potential for sub-THz communication and high-resolution radar systems.

Table 1. Performance comparison and results summary.

	This Work	[12]	[13]	[14]	[15]
Technology	130 nm SiGe	130 nm SiGe	130 nm SiGe	22 nm SOI	22 nm SOI
f_t/f_{max} [GHz]	300/450	250/370	230/280	240/230	240/230
f_{3dB}[GHz]	150–170	114–146	112–140	125–145	140–147
LO generation	X9 multiplier chain	Off-chip	X6 multiplier chain	X9 multiplier chain	X3 multiplier chain
P_{1dB} (input) [dBm]	−19 #	−7	−24	−30	−29
CG [dB]	28 #	25	27	27	27–30
NF [dB]	7.3 *	11.8 *	10 *	8.5 *	9
P_{DC} [mW]	428	326	280	198	197.5
Area [mm^2]	1.04	0.1	1.54	1.44	3.1

* Simulated; # experimental result.

Author Contributions: Conceptualization, F.H. and Z.W.; methodology, F.H.; software, F.H., Y.D., Z.X., M.N. and Y.T.; validation, F.H.; formal analysis, F.H.; investigation, F.H.; resources, Z.W.; data curation, F.H.; writing—original draft preparation, F.H., Y.D., Z.X., Z.Z., Z.S. and K.W.; writing—review and editing, F.H., Q.X. and Z.W.; visualization, F.H.; supervision, Z.W.; project administration, F.H. and Z.W.; funding acquisition, Q.X. and Z.W. All authors have read and agreed to the published version of the manuscript.

Funding: This work was supported in part by the National Natural Science Foundation of China under Grant 62034002 and Grant 61874022.

Informed Consent Statement: Not applicable.

Data Availability Statement: The data that support the findings of this study are available within the article.

Acknowledgments: The authors would like to thank the anonymous reviewers and editor for their valuable comments which helped in improving this manuscript.

Conflicts of Interest: The authors declare no conflict of interest.

References

1. Chen, Z.; Ma, X.; Zhang, B.; Zhang, Y.; Niu, Z.; Kuang, N.; Li, S. A survey on terahertz communications. *China Commun.* **2019**, *16*, 1–35. [CrossRef]
2. Akyildiz, I.F.; Han, C.; Hu, Z.; Nie, S.; Jornet, J.M. Terahertz Band Communication: An Old Problem Revisited and Research Directions for the Next Decade. *IEEE Trans. Commun.* **2022**, *70*, 4250–4285. [CrossRef]
3. Guan, S.; Cheng, J.; Chang, S. Recent Progress of Terahertz Spatial Light Modulators: Materials, Principles and Applications. *Micromachines* **2022**, *13*, 1637. [CrossRef] [PubMed]
4. Lin, C.-H.; Cheng, Y.-H. A THz Waveguide Bandpass Filter Design Using an Artificial Neural Network. *Micromachines* **2022**, *13*, 841. [CrossRef]
5. Song, R.; Li, N. High Speed Terahertz Communication in the Space and Terrestrial Integrated Next Generation Wireless Communication Systems. In Proceedings of the 2020 13th UK-Europe-China Workshop on Millimetre-Waves and Terahertz Technologies (UCMMT), Online, 29–31 August 2020; pp. 1–3. [CrossRef]
6. Stanko, S.; Caris, M.; Wahlen, A.; Sommer, R.; Wilcke, J.; Leuther, A.; Tessmann, A. Millimeter resolution with radar at lower terahertz. In Proceedings of the 2013 14th International Radar Symposium (IRS), Dresden, Germany, 19–21 June 2013; pp. 235–238.
7. Kanno, A.; Sekine, N.; Uzawa, Y.; Hosako, I.; Kawanishi, T. 300-GHz versatile transceiver front-end for both communication and imaging. In Proceedings of the 2015 40th International Conference on Infrared, Millimeter, and Terahertz waves (IRMMW-THz), Hong Kong, China, 23–28 August 2015; pp. 1–2. [CrossRef]
8. Kang, S.; Thyagarajan, S.V.; Niknejad, A.M. A 240 GHz Fully Integrated Wideband QPSK Transmitter in 65 nm CMOS. *IEEE J. Solid-State Circuits* **2015**, *50*, 2256–2267. [CrossRef]
9. Wang, Z.; Chiang, P.Y.; Nazari, P.; Wang, C.C.; Chen, Z.; Heydari, P. A CMOS 210-GHz Fundamental Transceiver with OOK Modulation. *IEEE J. Solid-State Circuits* **2014**, *49*, 564–580. [CrossRef]
10. Heinemann, B.; Barth, R.; Bolze, D.; Drews, J.; Fischer, G.G.; Fox, A.; Fursenko, O.; Grabolla, T.; Haak, U.; Knoll, D.; et al. SiGe HBT technology with fT/fmax of 300GHz/500GHz and 2.0 ps CML gate delay. In Proceedings of the 2010 International Electron Devices Meeting, San Francisco, CA, USA, 6–8 December 2010; pp. 30.5.1–30.5.4. [CrossRef]
11. Chevalier, P.; Jungemann, C.; Lovblom, R.; Maneux, C.; Ostinelli, O.; Pawlak, A.; Rinaldi, N.; Rucker, H.; Wedel, G.; Zimmer, T.; et al. Si/SiGe:C and InP/GaAsSb Heterojunction Bipolar Transistors for THz Applications. *Proc. IEEE* **2017**, *105*, 1035–1050. [CrossRef]
12. Aguilar, E.; Issakov, V.; Weigel, R. Highly-Integrated <0.14mm2D -Band Receiver Front-Ends for Radar and Imaging Applications in a 130 nm SiGe BiCMOS Technology. In Proceedings of the 2019 IEEE 19th Topical Meeting on Silicon Monolithic Integrated Circuits in RF Systems (SiRF), Orlando, FL, USA, 20–23 January 2019; pp. 1–4. [CrossRef]
13. Carpenter, S.; Zirath, H.; He, Z.S.; Bao, M. A fully integrated D-band direct-conversion I/Q transmitter and receiver chipset in SiGe BiCMOS technology. *J. Commun. Netw.* **2021**, *23*, 73–82. [CrossRef]
14. Farid, A.A.; Simsek, A.; Ahmed, A.S.H.; Rodwell, M.J.W. A Broadband Direct Conversion Transmitter/Receiver at D-band Using CMOS 22nm FDSOI. In Proceedings of the 2019 IEEE Radio Frequency Integrated Circuits Symposium (RFIC), Boston, MA, USA, 2–4 June 2019; pp. 135–138. [CrossRef]
15. Wang, C.; Rebeiz, G. A 2-Channel 136–156 GHz Dual Down-Conversion I/Q Receiver with 30 dB Gain and 9.5 dB NF Using CMOS 22 nm FDSOI. In Proceedings of the 2021 IEEE Radio Frequency Integrated Circuits Symposium (RFIC), San Diego, CA, USA, 11–13 June 2021; pp. 219–222. [CrossRef]
16. Kim, J.; Buckwalter, J.F. Staggered Gain for 100+ GHz Broadband Amplifiers. *IEEE J. Solid-State Circuits* **2011**, *46*, 1123–1136. [CrossRef]
17. Jang, T.H.; Jung, K.P.; Kang, J.S.; Byeon, C.W.; Park, C.S. 120-GHz 8-Stage Broadband Amplifier with Quantitative Stagger Tuning Technique. *IEEE Trans. Circuits Syst. I: Regul. Pap.* **2020**, *67*, 785–796. [CrossRef]
18. Karakuzulu, A.; Eissa, M.H.; Kissinger, D.; Malignaggi, A. A Broadband 110–170-GHz Stagger-Tuned Power Amplifier With 13.5-dBm Psat in 130-nm SiGe. *IEEE Microw. Wirel. Compon. Lett.* **2021**, *31*, 56–59. [CrossRef]
19. Turkmen, E.; Burak, A.; Guner, A.; Kalyoncu, I.; Kaynak, M.; Gurbuz, Y. A SiGe HBT D-Band LNA With Butterworth Response and Noise Reduction Technique. *IEEE Microw. Wirel. Compon. Lett.* **2018**, *28*, 524–526. [CrossRef]
20. Turkmen, E.; Aksoyak, I.K.; Debski, W.; Winkler, W.; Ulusoy, A.Ç. A 225–265 GHz I-Q Receiver in 130-nm SiGe BiCMOS for FMCW Radar Applications. *IEEE Microw. Wirel. Compon. Lett.* **2022**, *32*, 899–902. [CrossRef]

21. Liu, X.; Luong, H.C. A 170-GHz 23.7% Tuning-Range CMOS Injection-Locked LO Generator with Third-Harmonic Enhancement. *IEEE Trans. Microw. Theory Tech.* **2020**, *68*, 2668–2678. [CrossRef]
22. Chen, A.C.; Pham, A.-V.; Leoni, R.E., III. A Novel Broadband Even-Mode Matching Network for Marchand Baluns. *IEEE Trans. Microw. Theory Tech.* **2009**, *57*, 2973–2980. [CrossRef]

micromachines

MDPI

Article

An Ultra-Low-Power Analog Multiplier–Divider Compatible with Digital Code for RRAM-Based Computing-in-Memory Macros

Yiming Yang [1], Shidong Lv [1], Xiaoran Li [1,2,*], Xinghua Wang [1,2,3], Qian Wang [1], Yiyang Yuan [4], Sen Liang [1] and Feng Zhang [4]

[1] School of Integrated Circuits and Electronics, Beijing Institute of Technology, Beijing 100081, China; yangyiming@bit.edu.cn (Y.Y.); 3220225265@bit.edu.cn (S.L.); 89811@bit.edu.cn (X.W.); 3120210707@bit.edu.cn (Q.W.)
[2] BIT Chongqing Institute of Microelectronics and Microsystems, Chongqing 401332, China
[3] Yangtze Delta Region Academy of Beijing Institute of Technology, Jiaxing 314000, China
[4] Institute of Microelectronics, Chinese Academy of Sciences, Beijing 100029, China; yuanyiyang@ime.ac.cn (Y.Y.); zhangfeng_ime@ime.ac.cn (F.Z.)
* Correspondence: xiaoran.li@bit.edu.cn

Abstract: This manuscript presents an ultra-low-power analog multiplier–divider compatible with digital code words, which is applicable to the integrated structure of resistive random-access memory (RRAM)-based computing-in-memory (CIM) macros. Current multiplication and division are accomplished by a current-mirror-based structure. Compared with digital dividers to achieve higher precision and operation speed, analog dividers present the advantages of a reduced power consumption and a simple circuit structure in lower precision operations, thus improving the energy efficiency. Designed and fabricated in a 55 nm CMOS process, the proposed work is capable of achieving 8-bit precision for analog current multiplication and division operations. Measurement results show that the signal delay is 1 µs when performing 8-bit operation, with a bandwidth of 1.4 MHz. The power consumption is less than 6.15 µW with a 1.2 V supply voltage. The proposed multiplier–divider can increase the operation capacity by dividing the input current and digital code while reducing the power consumption and complexity required by division, which can be further utilized in real-time operation of edge computing devices.

Keywords: CMOS; computing-in-memory; current mirror; edge computing; multiplier–divider

Citation: Yang, Y.; Lv, S.; Li, X.; Wang, X.; Wang, Q.; Yuan, Y.; Liang, S.; Zhang, F. An Ultra-Low-Power Analog Multiplier–Divider Compatible with Digital Code for RRAM-Based Computing-in-Memory Macros. *Micromachines* **2023**, *14*, 1482. https://doi.org/10.3390/mi14071482

Academic Editor: Niall Tait

Received: 7 June 2023
Revised: 17 July 2023
Accepted: 18 July 2023
Published: 24 July 2023

1. Introduction

Dividers are critical components in many applications. In algorithms that involve large-scale computation such as convolutional neural networks (CNN), division operations are required during normalization processes and other related algorithms. Intelligent edge computing equipment that involves image processing often requires several steps such as image acquisition, image pre-processing, neural network operation, and result output. In addition to the normalization process in the neural network algorithm that requires division, some image preprocessing algorithms also require division before the data are input into the neural network, such as histogram equalization, batch normalization (BN), mean filtering, etc.

Histograms are commonly used to represent the distribution of brightness levels in an image. By extending the histogram to a wider range, the contrast of the image can be easily enhanced, which in turn can improve the accuracy of network-based reasoning. This process involves integrating the pixel values for each brightness level and performing simultaneous multiplication and division by N to normalize the values. Figure 1 illustrates the effect and mathematical equations of the histogram equalization algorithm. Adding histogram equalization to the pre-processing stage can enhance overall accuracy [1]. The purpose of batch normalization (BN) is to standardize the output of the middle layer of the CNN. During the BN process, the sample mean and variance of the layer inputs are computed and adjusted. BN can stabilize and accelerate the training of the neural network and facilitate better gradient flow throughout the network. Mean filtering is a technique that involves smoothing by reducing noise and eliminating high-frequency components from an image. It operates on each pixel, multiplying and dividing its matrix by coefficients and summing the resulting products. This process requires simultaneous multiplication and division of an arbitrary and variable number. The weighted mean algorithm is shown in Figure 2. Mean filtering can effectively filter out additive Gaussian noise in the image and perform a mean pooling operation between each layer of the neural network, thereby improving its accuracy [2].

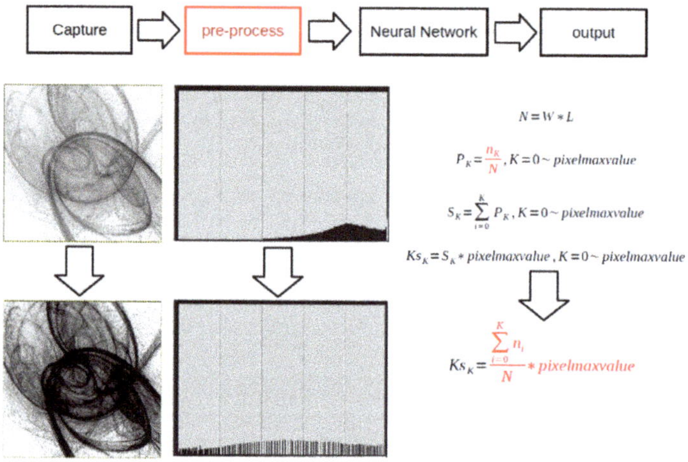

Figure 1. The effect and mathematical equations of the histogram equalization algorithm. Division is inevitably applied for the equation marked in red.

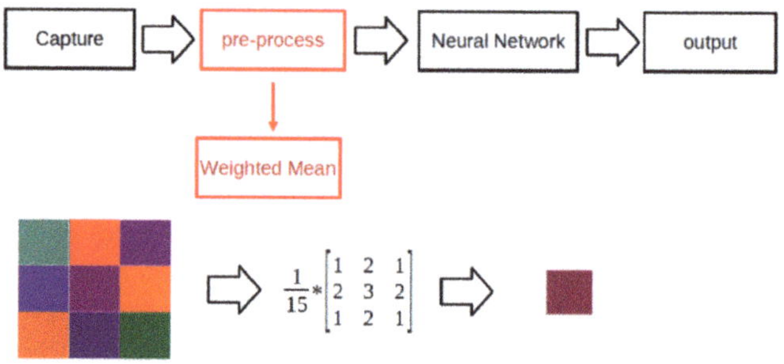

Figure 2. The weighted mean algorithm, which is another image preprocessing algorithm. Division is inevitably used in the mean process.

With the development of CNNs, the huge amount of complicated computation severely limits the application of CNNs in edge intelligent Internet of Things (IoT) devices due to the limitation of power consumption. Non-volatile computing-in-memory (nvCIM) architecture based on memristors for building up neural networks effectively breaks the bottleneck of the memory wall [3]. Resistive random-access memory (RRAM)-based CIM accelerators have significant advantages in edge computing due to the potential to perform CNN computations in an energy-efficient manner. RRAM can accelerate CNN computations and export the result of matrix–vector multiplication in an analog current form, making it a promising technology for edge devices. While multiplication and addition have been extensively researched for RRAM-based CIM, division has been mostly avoided due to the difficulty of implementation. Digital dividers can be utilized for computing, but additional analog-to-digital converters (ADC) are needed for data transmission. Furthermore, traditional digital division suffers from a higher power consumption, a larger area, and a lower speed. Analog dividers can achieve high speed while reducing the power consumption and circuit complexity [4].

Conventional analog multipliers and dividers can be mainly implemented through two approaches. One is to use current-mode amplifiers [5–8], such as current-following transconductance amplifiers (CFTA) [6], current-controlled conveyors (CCC) [7,8], etc. These active amplifiers provide a higher gain, which leads to more precise calculation results. However, these amplifiers may increase the power consumption and are commonly implemented using bipolar transistors that require a high supply voltage, generally requiring a negative supply voltage. Another kind of multiplier and divider is based on the unique square characteristic of MOSFET [9–12], or the exponential characteristic of the weak inversion region [13–16], which can achieve a higher bandwidth, smaller size and lower power consumption. However, due to the limited working area of the transistor, the operation range is narrow. In addition, the output is an approximate value, and the accuracy at both ends of the operation range is slightly worse. Recently, a new pulse-based analog divider was proposed, which can use a digital code word as output and is compatible with digital systems with high precision [17]. However, it suffers from a large area, a high power consumption, and a relatively slow speed. As CMOS technology continues to develop, the speed and accuracy of current mirrors are becoming higher, providing a possibility for the implementation of new analog multiplication and division circuits.

This manuscript proposes an analog multiplier–divider compatible with digital code words that can be used for real-time division operations of edge devices and to improve energy efficiency. The current can be conveniently and precisely multiplied or divided through the current mirror. The multiplication and division of the current can be completed only by controlling the multiples at both ends of the current mirror. The whole process requires no approximation and can satisfy a large range of inputs and outputs with a high accuracy. At the same time, only two transistors have continuous current in the whole circuit, resulting in an increased efficiency. The circuit size required to perform 8-bit operations is also small. Therefore, the proposed small size and ultra-low-power multiplier–divider is suitable for real-time operations in an RRAM-based CIM macro to improve the computing capacity.

The remainder of this paper is organized as follows. Section 2 describes the circuit implementation details of the proposed multiplier–divider. The measurement results are discussed in Section 3. A conclusion is provided in Section 4.

2. Circuit Implementation

The output current can be multiplied or divided by controlling the current mirror multiples. As shown in Figure 3, the output current can be calculated in Equation (1).

$$I_{out} = \frac{M}{D} \times I_{in} \tag{1}$$

It is difficult to change the ratio (D and M) of the current mirror after fabrication. Therefore, for practical use, it is necessary for both the divisor and dividend to be variable, allowing for continuous adjustment or at least covering all integer values.

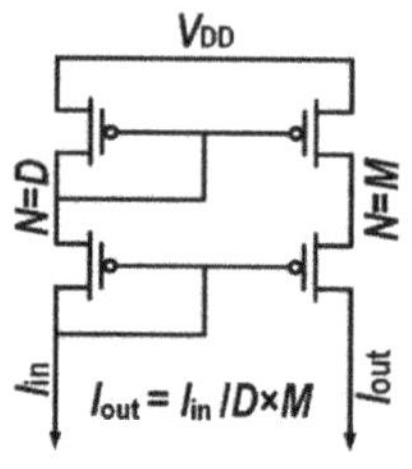

Figure 3. The basic structure of current multiplication and division using current mirror. The current multiples on the input side and output side are D and M, respectively.

Figure 4 shows the current mirror with any integer multiple. Only the transistor gate is connected to the circuit, reducing the load capacitance of the input stage. To ensure accuracy, a two-stage cascode current mirror is employed. The input current mirrors are divided into eight groups with multiples of 1, 2, 4, 8, 16, 32, 64, and 128, respectively. As a result, a total of 255 times input current can be connected, performing 8-bit operation. When $N = M_{IN} = 1$, the multiple of the output current mirror is 1. The output current is I_{in}, divided by the current mirror multiple on the input side. When the divisor input is high, the gates of the input-side current mirrors are connected to the output-side transistor gates and to the input-side transistor drains. When the divisor input is low, the gates will be connected to V_{DD}. This allows calculations of all 8-bit binary multiples, thus completing the division between the analog input current and the 8-bit divisor. When the input current is an integer multiple of the discrete amplitude of the unit current, the output current is the multiple of the operation result of the integer of the same unit current.

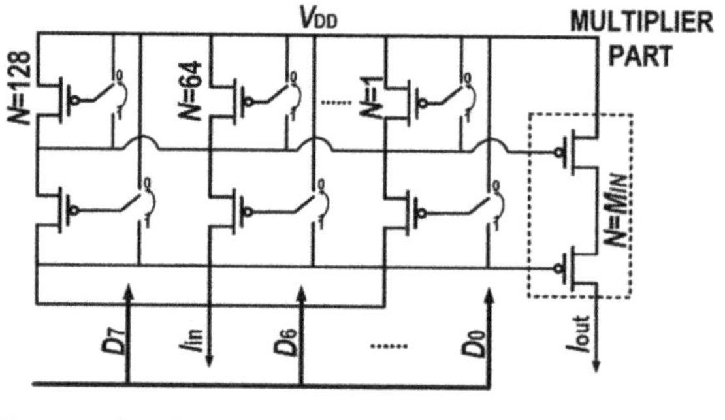

Figure 4. The schematic of the divider part. The divisor D can be changed from 0 to 255 by switches controlled by digital code D_0~D_7.

Meanwhile, the output is a continuous value since the proposed structure retains the characteristics of an analog circuit. In this case, the multiple will not be rounded to the decimal part, if any, which can meet the rounding requirements often required in image processing algorithms. The circuit is also naturally compatible with division by 0. When D_{IN} is zero, all the gates of the input-side transistors will be connected to V_{DD}, while the gates of the output-side transistors will be directly connected to the drains at the input side. There will be a large voltage drop due to the high turn-off resistance with current input. However, transistors farther from the power supply will have higher threshold voltages and higher turn-off resistance due to substrate bias effects for the cascode structure. As a result, most of the voltage drops will occur on this transistor, while transistors closer to the power supply will have less voltage drops. The gate of the transistor that is farther from the power supply on the output side is almost grounded and turned on, while the gate voltage of the transistor closer to the power supply is almost the power supply voltage. At this time, the total current is the current of the transistor with the smaller current, and the output is almost zero, which can simulate the effective bit return of 0 after the division overflow.

Both mentioned operations require one multiplication and one division. Therefore, completing the multiplication simultaneously will increase the working efficiency of the system. Multiplication can be performed by controlling the multiple of the current mirror at the output side. As shown in Figure 5, the multiplier part is similar to the divider part, which is also composed of cascode current mirrors. The 8-bit input M_{IN} controls the connection of eight groups of gates. When the multiplier input is high, the gates of the output-side current mirrors will be connected to the gates of the input-side transistors. When the multiplier input is low, the gates will be connected to V_{DD}. By changing the multiple of the current mirror at the output side, the result of the division from the divider part can be multiplied by the 8-bit M_{IN} input. The final output current can be calculated in Equation (2), enabling the multiplication and division to be performed in a single step.

$$I_{out} = \frac{M_{IN}}{D_{IN}} \times I_{in} \tag{2}$$

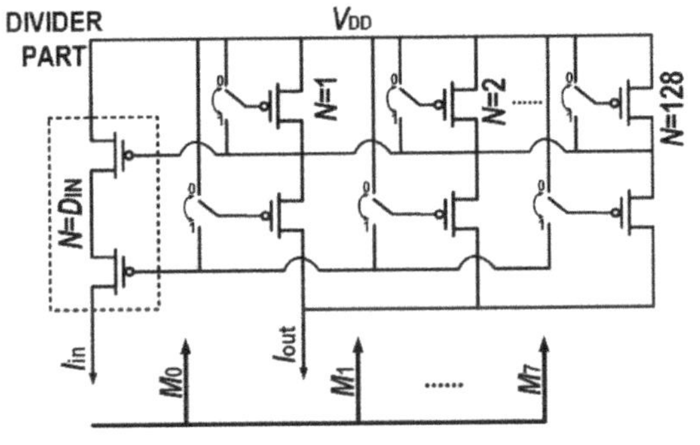

Figure 5. The schematic of the multiplier part. The multiple M can be changed from 0 to 255 by switches controlled by digital code M_0~M_7.

Since the multiplier part and the divider part have the same analog circuit characteristics, the output is the result of direct operation between the division result and the multiplier. Compared with the integer division followed by multiplication in digital circuits, the proposed work will not produce errors by discarding the decimal places directly. It can naturally guarantee the accuracy of the final result without any rounding during the operation, and if the result is also a decimal, the result will not be discarded. Similarly, when the multiplier input is 0, all the gates of transistors on the output side are connected to the power supply, that is, all the gates of transistors are turned off, which can ensure a very strong zero current output.

A schematic and the complete operation process of the proposed multiplier–divider are shown in Figure 6. The W/L of the single unit (when N = 1) PMOS in the current mirror is 4 μm/300 nm. When the input divisor is given at D_{IN} and the input multiplier is given at M_{IN}, the current mirrors on both sides are connected to D_{IN} and M_{IN}, respectively, while 255-D_{IN} and 255 min are disconnected, respectively. The input current flows through the current mirror to complete the multiplication and division operation. Figure 7 shows the parasitic capacitor connected during the operation. The input capacitance is (D_{IN} + M_{IN}) × C_{gs} and 255 × C_{ds}. Since only current operation is used, capacitor charging and discharging only require a relatively small voltage fluctuation, which can result in higher speeds and a lower power consumption than voltage operation. The output capacitance is 255 × C_{ds}. With the cascode current mirror load, the gain is relatively high with higher operation speeds.

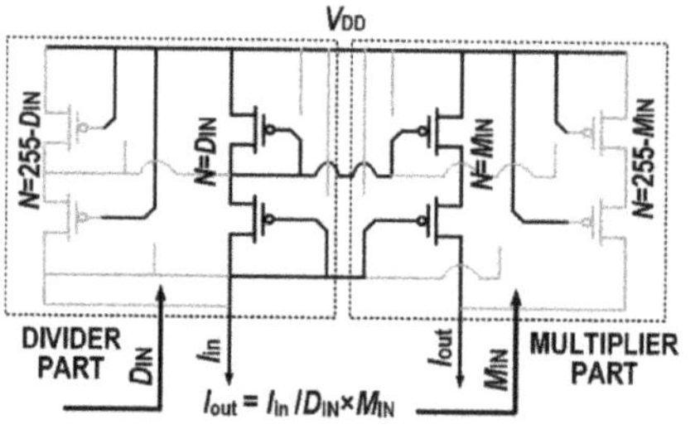

Figure 6. The schematic of the proposed multiplier–divider.

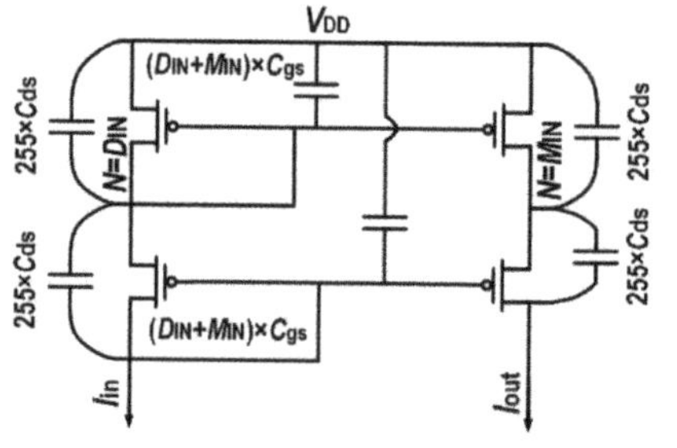

Figure 7. The parasitic capacitance of the divider.

Furthermore, the current mirror exhibits temperature-independent characteristics. Even if the temperature and threshold voltage change, as long as the transistors are matched, the operation precision will not be affected. In addition, the average current of the transistors on both sides of the current mirror is consistent, and the mismatch caused by serious unbalanced temperature distribution will not occur. Therefore, the accuracy only depends on the mismatch of the transistor caused by the deviation in fabrication, which is fixed after fabrication and can be calibrated.

The Monte Carlo simulation results shown in Figure 8 demonstrate the performance of the divider circuit. The simulation was conducted with a dividend of 255 (input current of 2550 nA), a divisor digital input of 2 (10 b), and a multiplier of 1. The deviation from the ideal value is only 0.31 pA, with an ideal value of 1.275 nA. The 2σ error was $\pm$193.7542 nA, which means that the error was within this range for more than 95% of the cases. These results show that the proposed circuit can provide accurate division with a high precision.

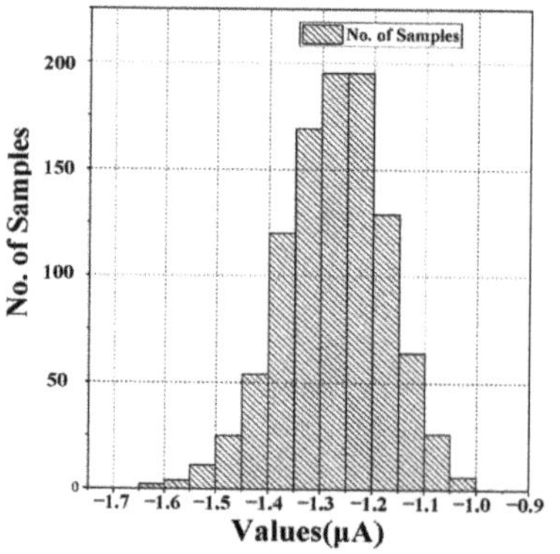

Figure 8. Monte Carlo simulation of the divider.

3. Measurement Results

Designed and implemented in a 55 nm ultra-low leakage CMOS process, the die micrograph of the proposed multiplier–divider is shown in Figure 9. The analog current operations enable the circuit to achieve 8-bit precision. The operating bandwidth is 1.4 MHz, and the signal delay for performing 8-bit operation is 1 μs. The maximum power consumption is lower than 6.15 μW under 1.2 V power supply.

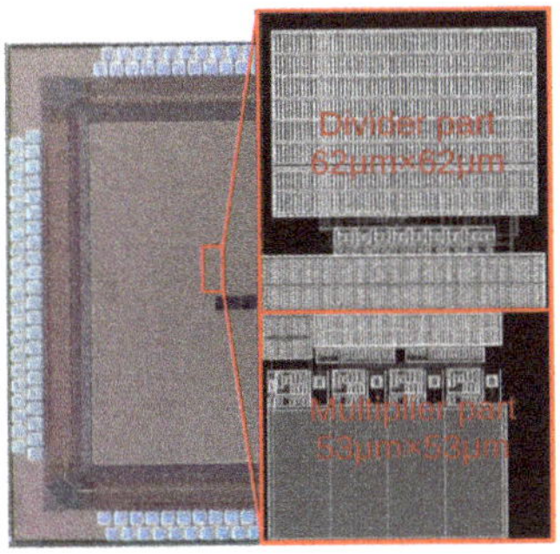

Figure 9. Die micrograph of the fabricated multiplier–divider.

The measurement results and ideal results of the analog divider are compared in Figure 10. The dividend used was binary 11111111, which is equivalent to decimal 255, and divisors ranging from 1 to 255 were used. The ideal result is the integer result after the ideal division is rounded, while the measurement result is the integer result corresponding to the analog current. When dividing by 0, the measured output is 0, and ideally the output will generate an error in the operation. The two curves show a slight divergence when the divisor is small, but as the divisor increased, the two curves become almost identical. Figure 11 shows the difference between the two curves in Figure 10. When the divisor is small, the absolute difference between the ideal and measured results can be up to 7. However, after the divisor reaches 25, the difference does not exceed 1.

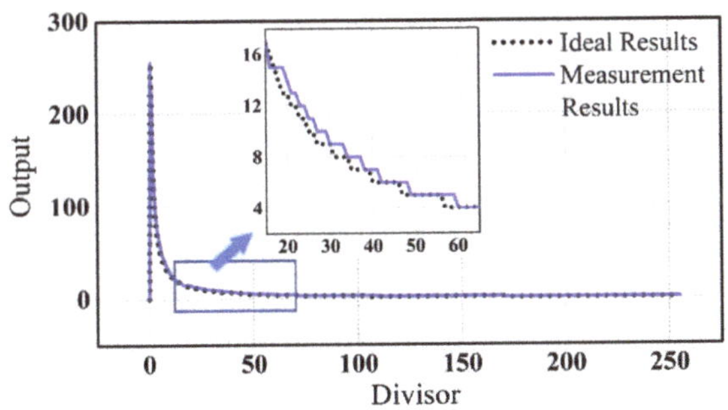

Figure 10. The measured output of the divider and the corresponding ideal division result. The dividend is 255. The x-axis represents the divisor, which is 0~255, and the y-axis is the division result. The measurement result is the current output after 8-bit analog to digital conversion. The LSB of the ADC is 10 nA and the ideal result is rounded after calculation.

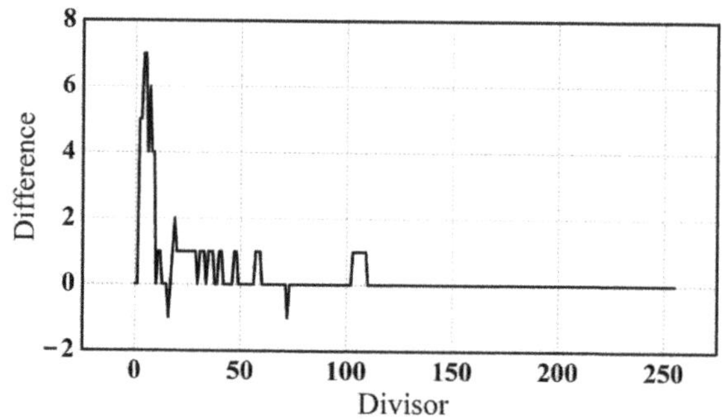

Figure 11. The subtraction between the ideal results and the measurement results of the divider.

Table 1 summarizes and compares the measurement performances of the implemented divider with previous state-of-the-art works. It can be found that the divider with CMOS characteristics can achieve higher speeds and a lower power consumption. The pulse-based divider in reference [17] has a higher output bit-width and is compatible with digital code words. However, it cannot perform multiplication, and suffers from a high power consumption and a large chip size. The resistor-based divider in reference [18] is also accurate. However, there is no advantage in power consumption and operation speed, since a separate amplifier chip is applied. Our proposed multiplier–divider exhibits the lowest peak power consumption, a relatively wide data range, and guaranteed, stable accuracy for all signals over the full data range, while is insensitive to temperature, making it compatible with digital code words.

Table 1. Performance summary with the state-of-the art works.

	[5]	[11]	[12]	[17]	[18]	This Work
Process	0.18 µm CMOS	0.5 µm CMOS	0.18 µm CMOS	0.35 µm CMOS	RRAM	55 nm CMOS
Working Mode	Current	Current	Current	Pulse	Voltage	Current
3 dB Bandwidth	59.7 MHz	18 MHz	62 MHz	/	/	1.4 MHz
Delay Time	/	/	/	/	0.5 ms	1 µs
Multiplier	Yes	Yes	Yes	No	Yes	Yes
Power Consumption	75 µW	120 µW	144 µW	<440 µW	/	<6.15 µW
Accurate or Approximate	Approximate	Approximate	Approximate	Accurate	Accurate	Accurate
Data Width or Range	1~10 µA	20 µA	1~20 µA	0.1~1 V, 12 bit	/	8-bit (256) (0~2550 nA)
Compatible with Digital	No	No	No	Yes	No	Yes
Chip Area	800 μm^2	0.021 mm^2	4050 μm^2	0.25 mm^2	/	6600 μm^2

4. Conclusions

An ultra-low-power, high-precision, and small-size analog multiplier–divider is presented in this manuscript. It is compatible with digital code words for RRAM-based computing-in-memory, and is capable of performing operations within the range of 8-bit integers. The fractional part of the middle process of multiplication and division can continue to participate in the operation. Designed and fabricated in a 55 nm CMOS process, the measurement results show that the maximum power consumption of the multiplier–divider is less than 6.15 µW under a 1.2 V power supply, and it can operate at a frequency of up to 1 MHz. The proposed work can offer an efficient and accurate solution for multiplication and division operations in various applications, which is suitable for real-time operation in

RRAM-based CIM macros, thus improving the computing capacity, and can be utilized in real-time operation of edge computing devices.

Author Contributions: Conceptualization, Y.Y. (Yiming Yang), X.W. and F.Z.; methodology, Y.Y. (Yiming Yang), S.L. (Shidong Lv) and S.L. (Sen Liang); software, Y.Y. (Yiming Yang) and Y.Y. (Yiyang Yuan); validation, Y.Y. (Yiming Yang), S.L. (Shidong Lv) and S.L. (Sen Liang); formal analysis, Y.Y. (Yiming Yang) and S.L. (Shidong Lv); investigation, Y.Y. (Yiming Yang), Q.W. and Y.Y. (Yiyang Yuan); resources, X.L., X.W. and F.Z.; data curation, Y.Y. (Yiming Yang); writing—original draft preparation, X.L. and Q.W.; writing—review and editing, X.L.; visualization, Y.Y. (Yiming Yang) and X.L.; supervision, X.L., X.W. and F.Z.; project administration, X.W. and F.Z.; funding acquisition, X.L., X.W. and F.Z. All authors have read and agreed to the published version of the manuscript.

Funding: This work was supported in part by the National Natural Science Foundation of China under Grant 62101038, in part by Beijing Institute of Technology Research Fund Program for Young Scholars under Grant XSQD-202105006, and in part by the Strategic Priority Research Program of Chinese Academy of Sciences, Grant No. XDA0330000 and Grand No. XDB44000000.

Data Availability Statement: All the data are contained within this article.

Conflicts of Interest: The authors declare no conflict of interest.

References

1. Mungra, D.; Agrawal, A.; Sharma, P. PRATIT: A CNN-based emotion recognition system using histogram equalization and data augmentation. *Multimed. Tools Appl.* **2020**, *79*, 2285–2307. [CrossRef]
2. Momeny, M.; Jahanbakhshi, A.; Jafarnezhad, K.; Zhang, Y.D. Accurate classification of cherry fruit using deep CNN based on hybrid pooling approach. *Postharvest Biol. Technol.* **2020**, *166*, 111204–111212. [CrossRef]
3. Huo, Q.; Yang, Y.; Wang, Y.; Lei, D.; Fu, X.; Ren, Q.; Xu, X.; Luo, Q.; Xing, G.; Chen, C.; et al. A computing-in-memory macro based on three-dimensional resistive random-access memory. *Nat. Electron.* **2022**, *5*, 469–477. [CrossRef]
4. Oshima, T.; Nakamura, Y.; Yamawaki, T. Fast CMOS Analog Multiplier and Divider with Continuous-Time Inverter-Based Flash Digitizer. *IEEE Trans. Circuits Syst. II-Express Briefs* **2022**, *69*, 934–938. [CrossRef]
5. Popa, C. Improved Accuracy Current-Mode Multiplier Circuits with Applications in Analog Signal Processing. *IEEE Trans. Very Large Scale Integr. (VLSI) Syst.* **2014**, *22*, 443–447. [CrossRef]
6. Mongkolwai, P.; Tangsrirat, W. CFTA-Based Current Multiplier/Divider Circuit. In Proceedings of the 2011 International Symposium on Intelligent Signal Processing and Communications Systems (ISPACS), Chiang Mai, Thailand, 7–9 December 2011. [CrossRef]
7. Petchakit, W.; Kiranon, W.; Wardkien, P.; Petchakit, S. A Current-Mode CCCII-Based Analog Multiplier/Divider. In Proceedings of the ECTI-CON2010: The 2010 ECTI International Confernce on Electrical Engineering/Electronics, Computer, Telecommunications and Information Technology, Chiang Mai, Thailand, 19–21 May 2010.
8. Parveen, T. Electronically Tunable Current Mode Temperature Insensitive Active Only Analog Multiplier/Divider. In Proceedings of the 2018 3rd International Conference and Workshops on Recent Advances and Innovations in Engineering (ICRAIE), Jaipur, India, 22–25 November 2018. [CrossRef]
9. Demosthenous, A.; Panovic, M. Low-voltage MOS linear transconductor/squarer and four-quadrant multiplier for analog VLSI. *IEEE Trans. Circuits Syst. I Reg. Pap.* **2005**, *52*, 1721–1731. [CrossRef]
10. Sawigun, C.; Mahattanakul, J. A 1.5V Wide-Input Range High-Bandwidth CMOS Four-Quadrant Analog Multiplier. In Proceedings of the 2008 IEEE International Symposium on Circuits and Systems, Seattle, WA, USA, 18–21 May 2008. [CrossRef]
11. Lopez-Martin, A.J.; De La Cruz Bias, C.A.; Ramirez-Angulo, J.; Carvajal, R.G. Compact Low-Voltage CMOS Current-Mode Multiplier/Divider. In Proceedings of the 2010 IEEE International Symposium on Circuits and Systems (ISCAS), Paris, France, 30 May–2 June 2010. [CrossRef]
12. De La Cruz-Blas, C.A.; Thomas-Erviti, G.; Algueta-Miguel, J.M.; López-Martín, A. CMOS analogue current-mode multiplier/divider circuit operating in triode-saturation with bulk-driven techniques. *Integration* **2017**, *59*, 243–246. [CrossRef]
13. Cracan, A.; Bonteanu, G.; Bozomitu, R. A Weak-Inversion CMOS Analog Multiplier/Divider Circuit. In Proceedings of the 2018 IEEE 24th International Symposium for Design and Technology in Electronic Packaging (SIITME), Iasi, Romania, 25–28 October 2018. [CrossRef]
14. Pawarangkoon, P.; Sawigun, C. A Compact Bulk-Driven Four-Quadrant Analog Multiplier in Weak Inversion. In Proceedings of the 2018 IEEE Asia Pacific Conference on Circuits and Systems (APCCAS), Chengdu, China, 26–30 October 2018. [CrossRef]
15. Danesh, M.; Jayaraj, A.; Chandrasekaran, S.T.; Sanyal, A. Ultra-Low Power Analog Multiplier Based on Translinear Principle. In Proceedings of the 2019 IEEE International Symposium on Circuits and Systems (ISCAS), Sapporo, Japan, 26–29 May 2019. [CrossRef]
16. Vatalaro, M.; Moposita, T.; Strangio, S.; Trojman, L.; Vladimirescu, A.; Lanuzza, M.; Crupi, F. A Low-Voltage, Low-Power Reconfigurable Current-Mode Softmax Circuit for Analog Neural Networks. *Electronics* **2021**, *10*, 1004. [CrossRef]

17. Chen, K.; Chen, T.; Wei, C. Novel Pulse-Based Analog Divider with Digital Output. *IEEE Solid-State Circ. Lett.* **2019**, *3*, 21–24. [CrossRef]
18. Gomez, J.; Vourkas, I.; Abusleme, A.; Sirakoulis, G.C.; Rubio, A. Voltage Divider for Self-Limited Analog State Programing of Memristors. *IEEE Trans. Circuits Syst. II-Express Briefs* **2020**, *67*, 620–624. [CrossRef]

micromachines

Article

Electromagnetic Susceptibility Analysis of the Operational Amplifier to Conducted EMI Injected through the Power Supply Port

Peng Huang [1,2], Bing Li [1,2,*], Mengyuan Wei [1,2], Xuchun Hao [1,2], Xi Chen [1,2], Xiaozong Huang [3], Wei Huang [3], Shuling Zhou [3], Xiaokang Wen [3], Shuguo Xie [1,2] and Donglin Su [1,2,*]

1 School of Electronic and Information Engineering, Beihang University, Beijing 100191, China; penghuang@buaa.edu.cn (P.H.); xieshuguo@buaa.edu.cn (S.X.)
2 Research Institute for Frontier Science, Beihang University, Beijing 100191, China
3 No. 24 Research Institute, China Electronics Technology Group Corporation, Chongqing 404100, China; hxzchn@163.com (X.H.)
* Correspondence: bingli@buaa.edu.cn (B.L.); sdl@buaa.edu.cn (D.S.)

Abstract: Operational amplifiers (op-amps) are widely used in circuit systems. The increasing complexity of the power supply network has led to the susceptibility of the power supply port to electromagnetic interference (EMI) in circuit systems. Therefore, it is necessary to investigate the electromagnetic susceptibility (EMS) of op-amps at the power supply port. In this paper, we assessed the effect of EMI on the operational performance of op-amps through the power supply port by a bulk current injection (BCI) method. Firstly, we conducted the continuous sine wave into the power supply port by a current injection probe and measured the change in the offset voltage under EMI. Secondly, we proposed a new method of conducted susceptibility and obtained the susceptibility threshold regularities of the op-amps at the power supply port under the interference of different waveform signals. Our study provided conclusive evidence that EMI reduced the reliability of the op-amp by affecting the offset voltage of op-amps and demonstrated that the sensitivity type of op-amps was peak-sensitive at the power supply port. This study contributed to a deep understanding of the EMS mechanism and guided the design of electromagnetic compatibility (EMC) of op-amps.

Keywords: electromagnetic environmental effect; electromagnetic interference (EMI); electromagnetic susceptibility (EMS); operational amplifier (op-amp); susceptibility threshold; voltage follower

Citation: Huang, P.; Li, B.; Wei, M.; Hao, X.; Chen, X.; Huang, X.; Huang, W.; Zhou, S.; Wen, X.; Xie, S.; et al. Electromagnetic Susceptibility Analysis of the Operational Amplifier to Conducted EMI Injected through the Power Supply Port. *Micromachines* **2024**, *15*, 121. https://doi.org/10.3390/mi15010121

Academic Editor: Zhiming Chen

Received: 10 December 2023
Revised: 30 December 2023
Accepted: 9 January 2024
Published: 11 January 2024

1. Introduction

Recent advances in the field of integrated chip-manufacturing processes have fostered the ever-increasing need for the development of reliable and high-performance operational amplifiers (op-amps). Op-amps have become one of the most common electronic devices in the field of analog and control integrated circuits [1–3]. At the front end of the signal transmission link, the op-amp can sense and amplify the function of the input signal [4–7]. At the end of the signal transmission link, the op-amp enhances the drive capability of the system and performs impedance conversion. The architecture of the signal transmission link system is presented in Figure 1.

As chips become more integrated and data processing becomes faster, the electromagnetic interference (EMI) of chips must be methodically considered [8–10]. In addition, integrated circuits face more complex electromagnetic environments [3,11]. Op-amps as analog integrated chips are more susceptible to EMI [12,13]. The reliability of the op-amps directly affects the reliability of the whole circuit system. Therefore, some investigations have attracted more concern to reduce and improve the reliability of op-amps. For instance, 4H-SiC fabrication technology enables op-amps to operate at high temperatures with reliable performance [14,15]. An op-amp architecture that reduces the EMI effects has less EMI-induced offset [16–18]. Pseudo-differential inverters are commonly utilized to lessen the common-mode gain of the op-amps and enhance the differential mode gain to improve the common-mode rejection ratio of the op-amps [19]. The op-amp with symmetric topologies is capable of increasing immunity against EMI [20]. Due to the variety of interference sources in the electromagnetic environment, it is usually difficult to establish a very accurate electromagnetic environment model [21–25].

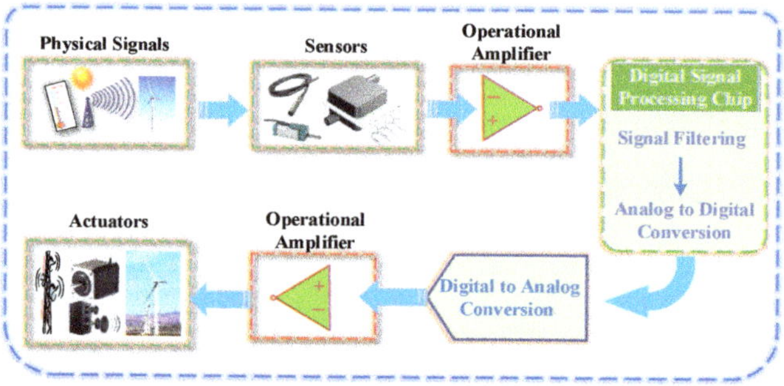

Figure 1. Signal transmission link system.

However, integrated circuits have many internal logic circuits and buffers. Their state switching can cause significant voltage drops and ripples in the distribution network, and supply voltage fluctuations can lead to severe EMC problems [26,27]. Prior research has neglected the impact of interference signal waveforms on the EMS properties of operational amplifiers. In addition, most people were concerned about the effect of injecting interference signals at differential inputs on op-amps, while injecting interference into power supply ports is lacking [28]. Furthermore, most research on power supply ports has only examined electrostatic discharges and low-frequency industrial-frequency signal interference, neglecting to address the impact of high-frequency signals on power supply ports. Moreover, the relationship between the electromagnetic susceptibility of op-amps and the type of interfering signals has yet to be investigated. We analyzed the effect of the type of interfering signal on the susceptibility threshold regularity of op-amps from several typical interfering signal waveforms.

In this paper, we focused on the EM characteristics of op-amps to conduct electromagnetic interference injected through the power supply port. Firstly, the continuous sine wave was injected into the power supply port, and the reliability of the op-amp reduced as the offset voltage increased significantly when the interference signal intensity exceeded the susceptibility threshold. Secondly, to obtain the comprehensive EMS of op-amps, five different waveforms were injected into the power supply port, respectively, and it revealed that the stability of the EMS threshold was essentially related only to the peak value and frequency of the interfering signal, not the duty cycle and bandwidth. The findings greatly contributed to the design of an electromagnetic protection of the op-amp.

The innovations of this paper can be summarized as four major aspects: (i) It was confirmed that EMI through the power port affected the op-amp's performance parameters; (ii) In the conducted susceptibility tests, we employed several different interference signals from the interference in the two standards MIL-STD-461G and GJB 151B-2013 to investigate the electromagnetic susceptibility characteristics of the op-amps [29,30]; (iii) Electromagnetic susceptibility threshold regularities of op-amps were found under signal interference of different waveforms; (iv) We discovered that the op-amp's susceptibility type was a peak type, and that the duty cycle and offset frequency of the modulated signal had little effect on the electromagnetic susceptibility thresholds of op-amps.

The rest of this article is structured as follows. In Section 2, we analyze the internal circuit characteristics of op-amps. Subsequently, we investigate the impact of EMI on the offset voltage of op-amps. Additionally, we conducted experiments to assess the susceptibility of op-amps to conducted EMI. In Section 3, we analyze the EMS characteristics under various waveform signals. In Section 4, we analyze the susceptibility and the difference between the proposed and the traditional test methods. Finally, we deliver the major conclusions.

2. Device and Experiment

2.1. Characteristics of Operational Amplifier Devices

The op-amp model SF158MD was used to examine the EMS features of the op-amp. The op-amp was a monolithic integrated circuit fabricated using a bipolar process. Furthermore, the SF158MD chip internally consisted of two independently operating low-power op-amps. Considering that the dual op-amps inside the SF158MD chip were manufactured using the same craft, we believed that the dual op-amps had the same susceptibility characteristics, so we selected one of the op-amps as the object of investigation. Figure 2a depicts this op-amp's pin arrangement and function. The op-amp's internal circuit consisted of four main circuit modules: differential input stage, intermediate gain stage, buffer output stage, and bias circuit. The op-amp's functional block diagram is shown in Figure 2b.

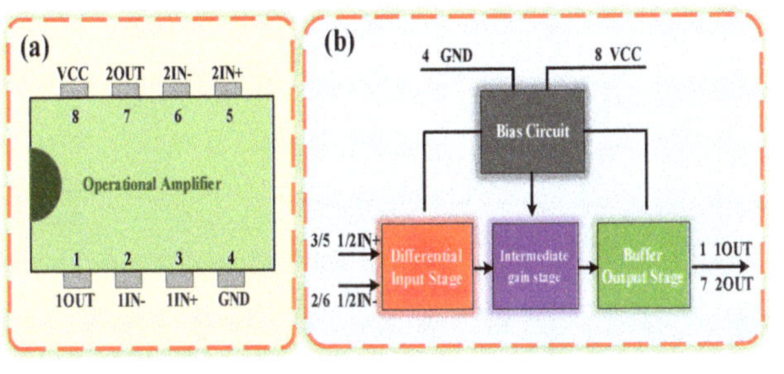

Figure 2. (**a**) Op-amp's pin configuration and function. (**b**) The operational amplifier's functional block diagram.

The SF158MD we used is the same type of operation amplifier as the LM158, LM358B, and LM358BA devices. By referencing the chip manuals, we found they had the same functional structure with similar differential input circuits, current mirror circuits, intermediate stage amplifier circuits, bias circuits, and buffer output circuits. Therefore, we considered that their simplified op-amp circuits are similar. Figure 3 shows the simplified op-amp circuit diagram. The purpose of the current mirror in the input stage was to supply an arbitrary DC bias current to the integrated circuit (IC). Simultaneously, the current mirror ensured that the output current I_{out1} was constant. The function of the differential input stage was to amplify differential signals and convert them into single-ended signals. The formula for calculating the PN junction currents of the two transistors passing through the current mirror is as follows.

$$\begin{cases} I_{c3} = I_{s3}\left(e\dfrac{V_{be3}}{V_T} - 1\right) \\ I_{c4} = I_{s4}\left(e\dfrac{V_{be4}}{V_T} - 1\right) \\ V_T = \dfrac{KT}{q} \end{cases} \tag{1}$$

In the differential input stage, two pairs of matched transistors, Q1–Q2 and Q3–Q4, constituted a symmetrical circuit structure. I_{c1}, I_{c2}, I_{c3}, and I_{c4} were the collector currents of the differential input stage transistors Q1, Q2, Q3, and Q4, respectively. I_{b3} and I_{b4} corresponded to the base currents of Q3 and Q4, respectively. Moreover, transistors Q3 and Q4 constituted a current mirror. A transistor had two PN junctions, of which I_s was the reverse saturation current of the PN junctions, V_{be} was the applied voltage across the PN junction, V_T was the voltage equivalent of the temperature, K was Boltzmann's constant, T was the temperature in Kelvin scale, and q was the charge of the electron. Since Q3 and Q4 were matched, $V_{be3} = V_{be4}$, and $I_{s3} = I_{s4}$. As shown below, we can obtain the relationship between the two currents, I_{c3} and I_{c4}.

$$I_{c3} = I_{c4} \tag{2}$$

We may construct the relationship equation for I_{c1}, I_{b4}, I_{b3}, and I_{b4} using Kirchhoff's Current Law (KCL), as shown below.

$$\begin{cases} I_{c1} = I_{b3} + I_{b4} + I_{c4} \\ I_{b3} = \dfrac{I_{c3}}{\beta_3} \\ I_{b4} = \dfrac{I_{c4}}{\beta_4} \end{cases} \tag{3}$$

β_3 and β_4 were the current amplification factors of transistors Q3 and Q4, respectively. Under the assumption that these two triodes were matched, β_3 and β_4 were the equal $\beta_3 = \beta_4 = \beta$. Formula (4) can be derived from Formulas (2) and (3).

$$I_{c1} = I_{c4}\left(\frac{2}{\beta} + 1\right) \tag{4}$$

If $\beta >> 2$, we can obtain $I_{c1} = I_{c3} = I_{c4}$. This ensures that the current mirror provides a consistent output current I_{out1}, which keeps the op-amp operating at a stable performance.

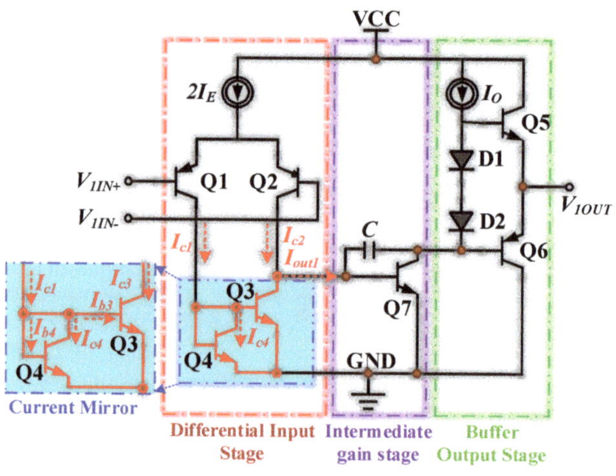

Figure 3. The simplified op-amp circuit diagram.

2.2. Op-Amp Offset Voltage Test under EMI

The offset voltage was a crucial op-amp performance indicator of the op-amp and directly determined the operating state of the op-amp. The voltage follower circuit was chosen to analyze the effect of EMI at the power supply port on the op-amp's offset voltage. The op-amp of the voltage follower acted as a buffer and did not amplify the signal. Moreover, the output voltage of the voltage follower was the same as the input voltage. The influence of EMI on op-amps was more easily detected by a voltage follower composed of op-amps. The op-amp's offset voltage test circuit under EMI is shown in Figure 4a. We connected the non-inverting input port of the op-amp to the ground while monitoring the input voltage values V_{1IN+} and the output voltage values V_{1OUT}. Under ideal conditions, the input and output voltages of the follower were both 0 V. The inverting input of the test circuit was directly connected to the output port. R_F and R_I represented the internal impedances of the wiring, which were small enough to be ignored. The block diagram of the op-amp offset voltage test is shown in Figure 4b. The experimental block diagram comprised the following components: digital multimeter, spectrum analyzer, radio frequency (RF) signal generator, RF power amplifier, DC power supply, ferrite core, bulk current injection probe, monitor probe, SF158MD op-amp, and waveform generator.

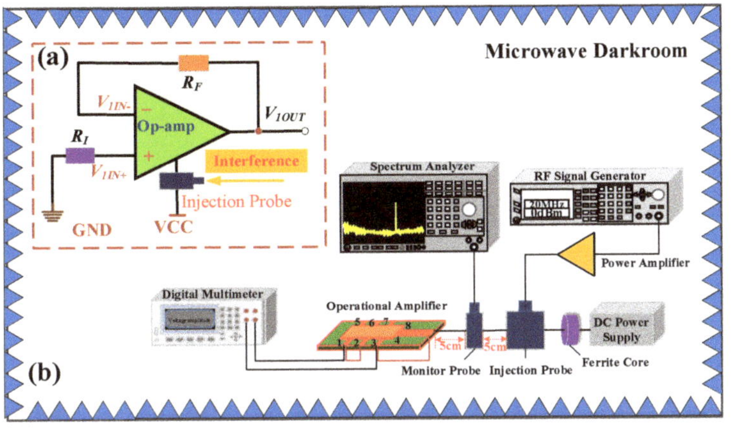

Figure 4. (**a**) Op-amp's offset voltage test circuit under EMI. (**b**) Block diagram of the op-amp's offset voltage test.

2.3. Op-Amp Conducted Electromagnetic Susceptibility Test

To investigate the EMS characteristics of op-amps during operation, we devised a voltage follower circuit to analyze the op-amp's electromagnetic susceptibility (EMS). The circuit schematic is depicted in Figure 5a. The input signal V_{in}, produced by the waveform generator was used as the operational input signal for the op-amp. The inverting input of the test circuit was directly connected to the output port. R_F' and R_I' represented the internal impedances of the wiring, and the resistance value was small and negligible. The interference signals were injected into the op-amp's power supply port through the injection probe. As long as the interference strength was below the EMS threshold, the op-amp would remain functioning correctly, and the input signal's voltage waveform would be identical to the output signal's voltage waveform, as depicted in Figure 5b. Conversely, if the intensity of the interference surpassed the threshold of EMS, it would result in the op-amp functioning abnormally. The op-amp's abnormal operation increased the peak-to-peak voltage waveform of the output signal, which differed significantly from the voltage waveform of the input signal, as depicted in Figure 5c. The peak-to-peak voltage of the input and output waveforms of the op-amp was 1.5 V when there was no EMI. We stipulated that when the peak-to-peak voltage of the output waveform fluctuated up and down by more than 0.5 V, the op-amp was considered to be in an electromagnetic susceptibility. It meant that when the op-amp was electromagnetically susceptible, the peak-to-peak voltage of the output waveform of the op-amp was more than 2 V or less than 1 V. Figure 5c shows the output waveform of the op-amp under abnormal operating conditions caused by EMI. Under the influence of electromagnetic interference, the output waveform jitter changed and was no longer the same as the waveform in Figure 5b, and the peak-to-peak voltage of the output signal waveform greatly exceeded the proposed 2 V. At this point, the op-amp was in an electromagnetically sensitive state.

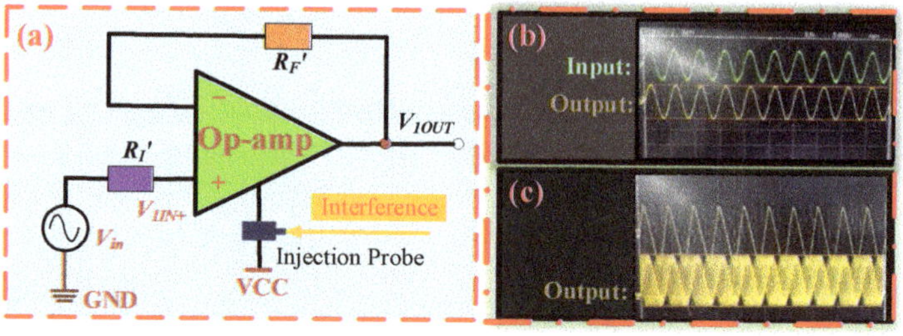

Figure 5. (**a**) Test circuit schematic for op amp in working condition under EMI. (**b**) The input and output waveforms of the op-amp under normal operating conditions. (**c**) The output waveform of the op-amp under abnormal operating conditions caused by EMI.

In the op-amp EMS test, several different signal waveforms from the RF source were injected into the power line of the op-amp by a bulk current injection probe. When the interfering signal's strength increased, the output waveform's peak-to-peak voltage exceeded 0.2 times the peak-to-peak voltage of the input signal. This indicated that the interfering signal achieved the susceptibility threshold of the operational amplifier at the frequency point. The conduction susceptibility tests utilized single-frequency continuous wave signals, pulse-modulated (PM) signals, and frequency-modulation (FM) signals as the interference signals for the op-amp. Table 1 displays the various types and characteristics of the interference signals. The continuous wave was a sinusoidal signal that spanned from 10 kHz to 400 MHz, representing an analog signal. Pulse-modulated signals exhibited varying duty cycles and a pulse frequency of 1 kHz, which differed from the standards presented in IEC 62132, MIL-STD-461G, and GJB 151B-2013 standards. Our test methods were based on all three of these test standards. The FM signals were sinusoidal signals modulated with variable frequency offsets. The fundamental waveform was a 10 kHz sinusoidal signal, which represented conventional linear FM radar signals. The carrier frequency range of pulse-modulated and FM signals was from 10 kHz to 400 MHz.

Table 1. Types and characteristics of the interference signals.

Interference Signal Types	Interference Signal Characterizations
Continuous wave	Continuous sine wave Frequency band: 10 kHz–400 MHz
Pulse-modulated signals	Duty cycles:10–90% Carrier frequency range: 10 kHz–400 MHz Fundamental waveform: pulse signals
Frequency-modulation signals	Frequency offsets: 1 kHz–10 kHz Carrier frequency range: 10 kHz–400 MHz Fundamental waveform: sinusoidal signals

The block diagram of the EMS test that was performed on the operational amplifier in this investigation is shown in Figure 6. The experimental block diagram comprised the following components: oscilloscope, spectrum analyzer, radio frequency (RF) signal generator, RF power amplifier, DC power supply, ferrite core, bulk current injection probe, monitor probe, SF158MD operational amplifier, and waveform generator. The waveform generator supplied a sinusoidal signal with a frequency of 200 kHz and a peak-to-peak voltage of 1 V to the input port of the op-amp as its functional signal. The oscilloscope observed the input signal waveform and output signal waveform of the op-amp in real-time. The RF signal generator generated various interference signals, which were injected into the operational amplifier's power supply wires by a bulk current injection probe. Both the injection probes and monitoring probes we used work efficiently in the 10 kHz–400 MHz band. Concurrently, the strength of the interference signal linked to the op-amp supply wire was monitored using a spectrum analyzer and a current monitoring probe. The DC power supply supplied 15 V to the op-amp. The ferrite core was shielded against EMI for the DC power supply. The operational amplifier's electromagnetic susceptibility test platform and the experimental instrumentation's connection are shown in Figure 7a, and the test circuit board is depicted in Figure 7b.

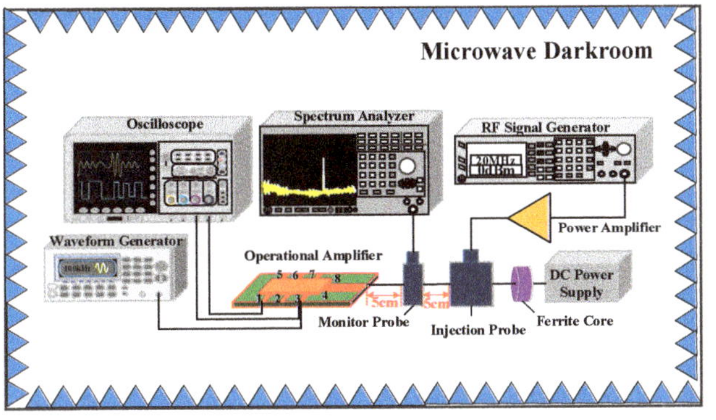

Figure 6. Block diagram of electromagnetic conducted susceptibility test of the op-amp.

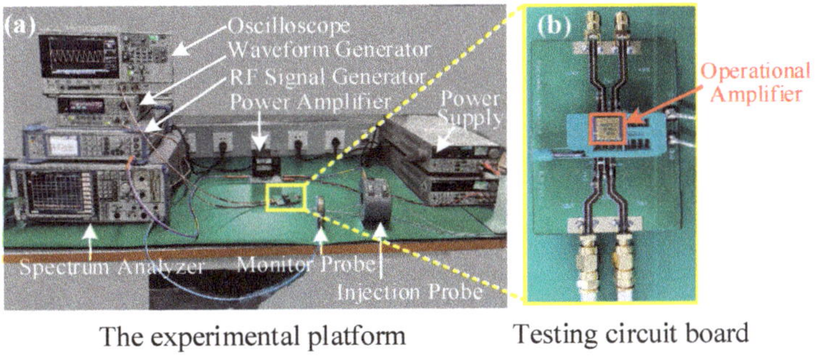

Figure 7. (**a**) Experimental platform for electromagnetic conduction susceptibility of the operational amplifier. (**b**) Testing circuit board.

The EMS test of the op-amp was conducted in a microwave darkroom to prevent electromagnetic interference from the external environment and maintain the reliability of the experimental process. The op-amp's EMS test's preparation stage referred to MIL-STD-461G and GJB 151B-2013. The EMS test procedure for op-amps was as follows.

(1) Preparation of the chip under test, the measuring instrument and the cables, and connection of the op-amp to the measuring device according to Figure 6. The current monitoring probe was placed 5 cm from the op-amp power supply port. The bulk current injection probe was positioned at a distance of 5 cm from the monitoring probe.
(2) Configuration of the test platform: Microwave-absorbing materials were placed around the test platform to reduce the impact of space electromagnetic fields on the test and ensure the experiment's reliability. We arranged the site of the experiment in a microwave darkroom. The power supply could provide a stable power supply to the operational amplifier and experimental equipment to ensure accurate measurements of the operational amplifier and the measuring instrument.
(3) Calibration of the measuring equipment: We calibrated the bulk-current injection probes, proper functioning of the operational amplifiers, and signal generator. The signal generator was capable of providing several proposed interference signals.
(4) The susceptibility test of op-amps: Firstly, we increased the intensity of the interference signal gradually by the bulk current injection probe until the op-amp appeared to be a

susceptible phenomenon. Secondly, we recorded the susceptibility phenomenon and the input power of the interference signal.

(5) Experimental data processing and susceptibility analysis: We recorded the strength of the interference signal during susceptibility testing, established susceptibility threshold curves, and investigated its EMS properties.

3. Results

3.1. Offset Voltage Test Results

The voltage values of the input and output ports of the voltage follower were set as 0.02 mV and 0.6 mV, respectively, in the absence of the EMI signal. The voltage $\Delta V = | V_{1out} - V_{1IN+} |$ was the output offset voltage of the op-amp. Figure 8 depicts the EMI effect on the offset voltage of the op-amp. In the experiments, a 380 MHz interference signal was coupled to the connecting cable of the op-amp power supply port through the injection probe. When the interfering signal intensity was less than 28 dBm, the input and output signal voltages and the op-amp offset voltage were near 0 V. However, when the intensity of the interference signal exceeded 28 dBm, the offset voltage was substantially raised, which reduced the reliability of the op-amp. An op-amp with the EMI could produce an offset voltage of more than 200 mV, which was significantly different from normal operating conditions. The work confirmed that EMI at the op-amp's power port affects its performance parameters.

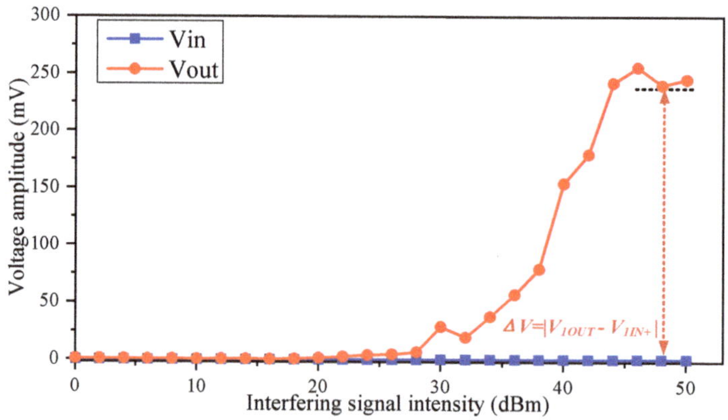

Figure 8. The effect of interfering signal intensity on offset voltage.

3.2. Conducted Susceptibility Test Results

A continuous sine wave with a frequency of from 10 kHz to 400 MHz was injected into the power port of the op-amp as an interference signal. There were differences in susceptibility thresholds at different frequency points. The op-amp was hardly susceptible to it when injecting interfering signals with frequencies below 10 MHz. When the frequency of the interfering signal was around 230 MHz and in the high-frequency band from 310 MHz to 400 MHz, it was more likely to make the op-amp electromagnetically susceptible. Under the current test conditions, the op-amp was not electromagnetic susceptible when the frequency of the interfering signal was lower than 10 MHz, so the susceptibility thresholds in the range from 10 kHz to 10 MHz were not shown on the threshold curves. On the contrary, the frequency bands in which the op-amps became electromagnetically susceptible were able to be shown on the threshold curves. The EMS threshold curve of the op-amp in the presence of sinusoidal signal interference is demonstrated in Figure 9.

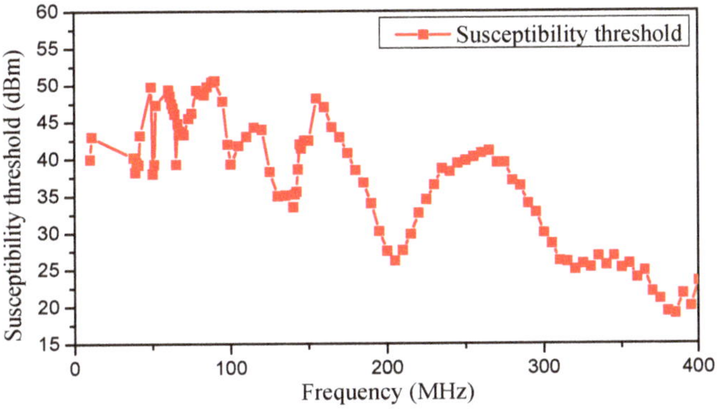

Figure 9. The EMS threshold curve of the op-amp subjected sinusoidal signal interference.

The pulse-modulated signals with a pulse period of 1 ms, and duty cycles of 10% and 50% were considered interference signals. Figure 10 illustrates the electromagnetic susceptibility threshold from 10 kHz to 400 MHz. The op-amp was hardly susceptible to it when the frequency of interfering signals was below 50 MHz. The EMS threshold curves subjected to pulse modulation (PM) signal interference were similar to the previously mentioned continuous sine wave signal interference. All of them had lower susceptibility thresholds near the 200 MHz frequency point and within the frequency range from 310 MHz to 400 MHz. The EMS threshold curves of the op-amp under pulse-modulated interference signals with different duty cycles are presented in Figure 10a. The sinusoidal continuous wave signal was regarded as the basic interference signal, and the single frequency point was selected based on the susceptibility threshold curve when the sinusoidal continuous wave was used as the interference signal. The minimum value of the electromagnetic susceptibility threshold for sinusoidal signal interference received by the op-amp corresponded to a frequency point of 380 MHz, which means that this frequency point was the most susceptible to the effects of EMI. Therefore, we chose this particular 380 MHz frequency point to analyze the electromagnetic susceptibility characteristics of the op-amp.

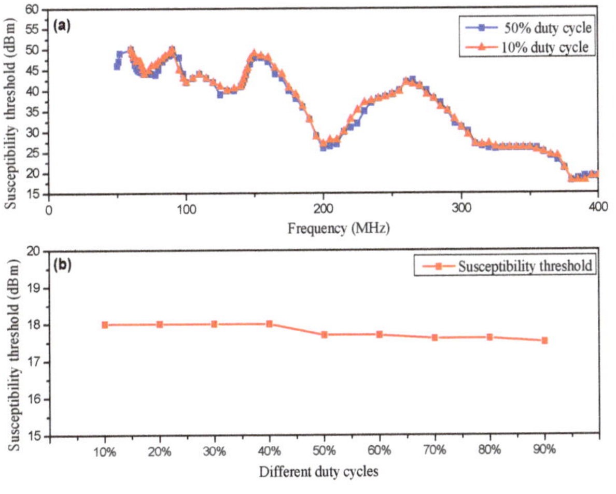

Figure 10. (**a**) The EMS threshold curves of the op-amps under pulse-modulated interference signals with various duty cycles. (**b**) EMS threshold curve of the op-amp under pulse-modulated interference signals with various duty cycles at 380 MHz.

We performed EMS tests at a fixed carrier frequency of 380 MHz and selected pulse-modulated signals with a duty cycle of from 10% to 90% in 10% steps as the interference signals. According to the experimental results, the EMS thresholds of the op-amp remained constant under the interference of pulse-modulated signals with various duty cycles. The EMS threshold curve of the op-amp under pulse-modulated interference signals with different duty cycles at 380 MHz is presented in Figure 10b.

Figure 11a shows the EMS threshold curves of the op-amp under the interference of sinusoidal frequency modulation (FM) signals with a frequency offset of 1 kHz and 9 kHz, respectively. The op-amps were hardly susceptible to it when the frequency of interfering signals was below 40 MHz. The carrier frequencies of interference signals had harmonic components at 40 MHz, 80 MHz, 120 MHz, and 200 MHz. The threshold of op-amps had less susceptibility near these frequency points and was more susceptible to EMI. Figure 11b illustrates the EMS threshold curve of an op-amp when the sinusoidal FM signal with different frequency offsets and a fixed carrier frequency of 380 MHz were the interfering signals. The susceptibility threshold of the op-amp remained essentially constant under the interference of sinusoidal FM signals with different frequency offsets.

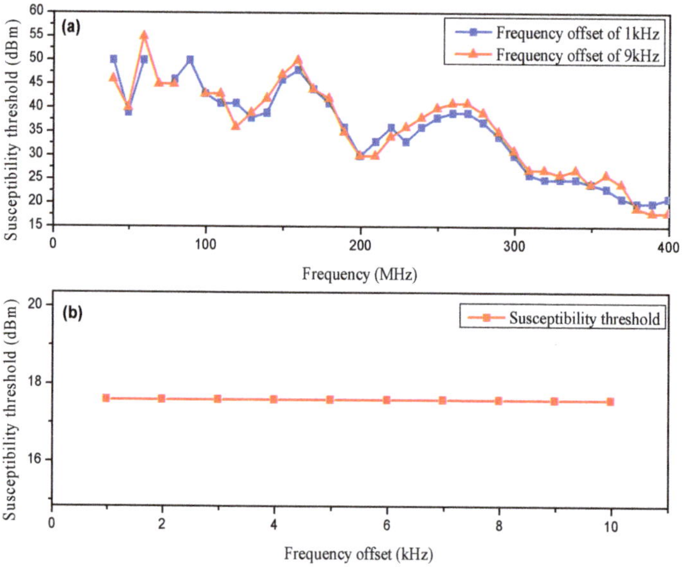

Figure 11. (**a**) The EMS threshold curves of the op-amp due to interference by sinusoidal frequency modulation signal interference with various frequency offsets. (**b**) EMS threshold curve of the op-amp due to interference by sinusoidal frequency modulation signal interference with different frequency offsets at 380 MHz.

Figure 12 shows that the monitored value measured by the monitor probe reflected the amount of energy coupled to the cable by the actual interference signal. The monitoring power can reflect the frequency selection characteristics of the op-amp, and the interference signal intensity of the RF source output can also reflect the susceptibility at different frequencies. The monitoring power and the output power of the RF source all reflect the electromagnetic susceptibility characteristics of the op-amp, so the output signal intensity of the RF source is chosen as the susceptibility threshold. We found that the monitoring values in the low-frequency band (<80 MHz) were almost all larger than −10 dBm, while the monitoring values in the high-frequency band (>230 MHz) were less than −20 dBm. Based on the comparison of the values of the monitoring probe, the interference signal in the low-frequency band may be coupled to the power supply cable of the op-amp more efficiently. However, the interference signal in the high-frequency band was more difficult to couple to the power supply cable of the op-amp. At the same time, we found that the op-amp had a higher susceptibility threshold at low frequencies and a lower susceptibility threshold at high-frequency bands. Although it was challenging for high-frequency interference signals to be coupled to the power supply cable of the op-amp, the op-amp was more susceptible to electromagnetic susceptibility from high-frequency interference signals, which could compromise their reliability.

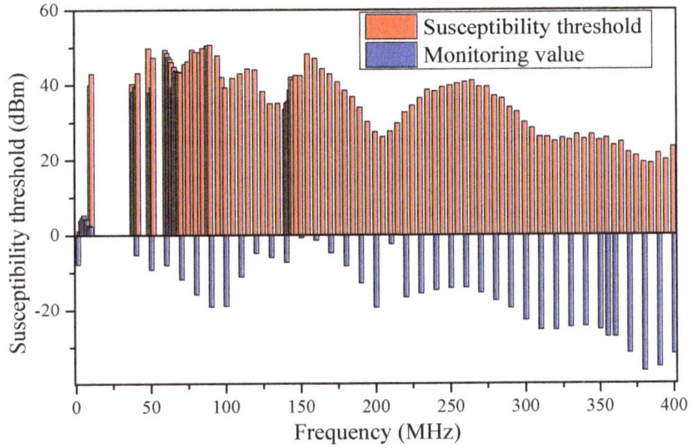

Figure 12. The EMS threshold thresholds in the presence of a single-frequency interference and monitoring values of monitoring probes.

4. Discussion

4.1. Susceptibility Analysis

As seen in the results in Section 3, we obtained the EMS threshold regularities for op-amp under interference signals with varying waveform parameters, such as frequency, modulation types, duty cycles, and frequency offsets. According to the experimental data in Figures 8–10, the op-amp had lower susceptibility thresholds to the high-frequency band signals, which meant that the op-amp was more easily affected by the high-frequency bands' interference. However, the interfering signals in the high-frequency band were more difficult to couple to the cable compared to the signals in the low-frequency bands in Figure 12. The reason why the op-amp operated abnormally was that the EMI signals coupled into the op-amp through the power supply cable and generated a DC shift [31]. EMI led to a non-linear reaction in the diode formed by the silicon PN junction in the op-amp. The non-linear response resulted in the production of a DC shift, as the interfering signals were rectified and converted into a DC signal.

Moreover, if the frequency of the high-frequency interference signals was much higher than the op-amp's bandwidth, the parasitic capacitances between the PN junction of the transistors inside the op-amp must be considered. The simplified op-amp circuit diagram under EMI is shown in Figure 13. i_{s1} and i_{s2} represented the currents coupled into the op-amp by the interfering signal. {(C_{cb1}, C_{be1}), (C_{cb2}, C_{be2}), (C_{cb3}, C_{be3}), (C_{cb4}, C_{be4}), (C_{cb5}, C_{be5}), (C_{cb6}, C_{be6}), (C_{cb7}, C_{be7})} represented the parasitic capacitance between the two PN junctions of {Q1, Q2, Q3, Q4, Q5, Q6, Q7} transistors, respectively. The parasitic capacitance disrupted the symmetrical balance of the differential amplification circuit, resulting in an unstable output from the current mirror. This instability ultimately produced a DC shift in the op-amp, negatively impacting its performance.

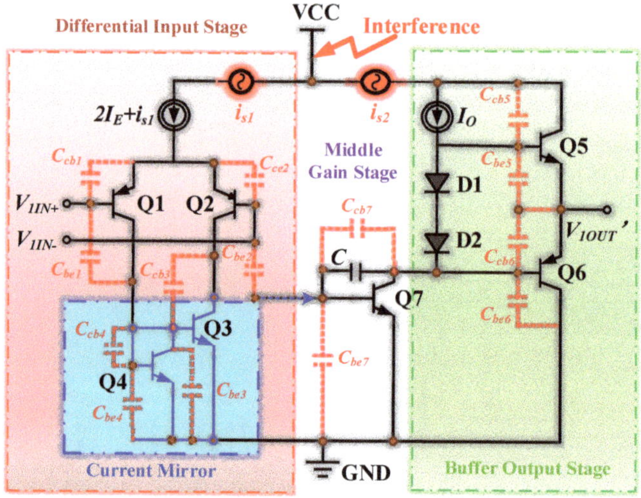

Figure 13. The simplified op-amp circuit diagram under EMI.

4.2. Differences in Test Methods

Within an intricate electromagnetic setting, all disruptive signals would ultimately impact the operational amplifier through the printed circuit board (PCB) traces or wire conduction, despite the electromagnetic interference being sent or emitted into the circuit system. Consequently, in this work, we proposed a novel BCI measurement methodology to investigate the susceptibility properties of the op-amp power supply port. According to IEC Standard 62132, various conventional methods for integrated chips mainly include direct power injection (DPI), a transverse electromagnetic (TEM) cell, a gigahertz transverse electromagnetic (GTEM) cell, and bulk current injection (BCI) [32,33]. TEM and GTEM are both radiated emission immunity tests, and they do not accurately analyze the EMS of the power supply ports of op-amps. BCI and DPI are both conducted emission immunity measurement methods. The DPI test method required additional design and processing of the test board, which was expensive. Moreover, the DPI test method cannot conduct tests in the chip's actual working environment. On the contrary, the BCI test method can directly carry out conducted susceptibility tests in the actual working environment of the chip without the need to design and process PCBs. Moreover, the BCI test was easy to operate. Standard BCI tests use a single pulse-modulated signal as the interference signal, so they cannot check how sensitive op-amps are to different electromagnetic environments. Compared to the traditional BCI method, the suggested method can obtain the electromagnetic susceptibility profiles of the op-amp in more dimensions by using signals with different waveforms as interference signals. There were no specialized cable ducts to fix the cables and to keep the cables 5 cm above the reference plane at the time of the test conditions. Throughout the test, to ensure that the test conditions were as consistent as possible, we did not stop the test and did not move the cables. However, it was hard to achieve repeatable susceptibility thresholds if repeat tests were conducted again after removing the test conditions. Because of the change in cable placement, the loop impedance was affected. This was also a shortcoming of this article. However, instead of focusing excessively on the frequency points of susceptibility and the magnitude of the susceptibility threshold, we paid more attention to the relationship between the waveform characteristics of different interfering signals and the susceptibility threshold curves. Therefore, this shortcoming had little effect on our investigation.

5. Conclusions

In this paper, we investigated the susceptibility properties of an op-amp through the power supply port. We injected interference signals by conduction at the power supply port to study the effects of EMI on the offset voltage. Moreover, we injected several different waveforms of interference signals into the power supply cables to investigate the effects of different waveforms on the electromagnetic susceptibility thresholds of the op-amp. We had shown that the op-amp's offset voltage changed dramatically when the interfering signal intensity exceeded 28 dBm. We found that the op-amp was not susceptible when the interfering signal frequency was less than 50 MHz. However, in the high-frequency band of from 300 MHz to 400 MHz, the susceptibility thresholds of the op-amps were all below 25 dBm and were more susceptible to EMS than in the low-frequency band. Moreover, the results revealed that the EMS of the op-amp was a type of "peak" susceptibility. The stability of the EMS threshold of the op-amp was essentially related only to the peak value and frequency of the interfering signal, not the duty cycle and bandwidth. This approach could obtain the EMS boundary of op-amps more comprehensively, which provides an effective method for evaluating the EMS characteristics of the op-amp. In addition, the comprehensive EMS boundary provides theoretical support for the electromagnetic protection of the circuit system.

Author Contributions: Conceptualization, P.H. and B.L.; methodology, P.H. and D.S.; software, B.L.; validation, M.W., X.H. (Xuchun Hao) and X.C.; formal analysis, P.H.; investigation, P.H.; resources, X.H. (Xiaozong Huang) and W.H.; data curation, S.Z. and X.W.; writing—original draft preparation, P.H.; writing—review and editing, P.H. and S.X.; visualization, P.H.; supervision, B.L. and D.S.; project administration, D.S.; funding acquisition, B.L. All authors have read and agreed to the published version of the manuscript.

Funding: This work was supported by the National Natural Science Foundation of China under Grant 62293492, Grant 62293495, and the Civil Aircraft Projects of China under Grant No. MJZ5-2N22.

Data Availability Statement: All the data are contained within this article.

Conflicts of Interest: Authors Xiaozong Huang, Wei Huang, Shuling Zhou and Xiaokang Wen were employed by the company No. 24 Research Institute, China Electronics Technology Group Corporation. The remaining authors declare that the research was conducted in the absence of any commercial or financial relationships that could be construed as a potential conflict of interest.

References

1. Katsu, S.; Kazumura, M.; Kano, G. Design and fabrication of a GaAs monolithic operational amplifier. *IEEE Trans. Electron Devices* **1988**, *35*, 831–838. [CrossRef]
2. Kularatna, N. (Ed.) Chapter 2—Operational Amplifiers. In *Modern Component Families and Circuit Block Design*; Newnes: Woburn, UK, 2000; pp. 47–85.
3. Gundla, J.; Boyapati, S.; Pasupureddi, V.S.R. Compact CMOS Miller OpAmp with High EMI-Immunity. *IEEE Trans. Electromagn. Compat.* **2020**, *62*, 2394–2400. [CrossRef]
4. Castoldi, A.; Guazzoni, C.; Naggi, A. Feasibility Study of the Use of Operational Amplifiers as forward Gain Stages in Charge Preamplifiers and Shaping Filters for Radiation Detectors. *IEEE Trans. Nucl. Sci.* **2023**, *70*, 1431–1435. [CrossRef]
5. Coudyzer, G.; Verplaetse, M.; Van Lombergen, B.; Borkowski, R.; Gurne, T.; Straub, M.; Lefevre, Y.; Ossieur, P.; Bonk, R.; Coomans, W.; et al. 100 Gbit/s PAM-4 Linear Burst-Mode Transimpedance Amplifier for Upstream Flexible Passive Optical Networks. *J. Lightwave Technol.* **2023**, *41*, 3652–3659. [CrossRef]
6. Ying, S.; Wang, L.; Zhao, Y.; Ma, M.; Ding, Q.; Xie, J.; Yao, D.; Mitra, S.; Chen, M.; Liu, T. A Novel In-Home Sleep Monitoring System Based on Fully Integrated Multichannel Front-End Chip and Its Multilevel Analyses. *IEEE J. Transl. Eng. Health Med.* **2023**, *11*, 211–222. [CrossRef]
7. Agouridis, D.C. Comments on "Noise performance of operational amplifier circuits" by F.N. Trofimenkoff and O.A. Onwuachi. *IEEE Trans. Educ.* **1992**, *35*, 98. [CrossRef]
8. Sicard, E.; Jianfei, W.; Shen, R.J.; Li, E.P.; Liu, E.X.; Kim, J.; Cho, J.; Swaminathan, M. Recent advances in electromagnetic compatibility of 3D-ICs—Part I. *IEEE Electromagn. Compat. Mag.* **2015**, *4*, 79–89. [CrossRef]
9. Sicard, E.; Jianfei, W.; Shen, R.; Li, E.P.; Liu, E.X.; Kim, J.; Cho, J.; Swaminathan, M. Recent advances in electromagnetic compatibility of 3D-ICs—Part II. *IEEE Electromagn. Compat. Mag.* **2016**, *5*, 65–74. [CrossRef]

10. Wang, L.; Chai, C.; Zhao, T.; Li, F.; Qin, Y.; Yang, Y. Mechanism Analysis and Multi-Scale Protection Design of GaN HEMT Induced by High-Power Electromagnetic Pulse. *Micromachines* **2022**, *13*, 1288. [CrossRef]
11. Walravens, C.; Winckel, S.V.; Redouté, J.M.; Steyaert, M. Efficient reduction of electromagnetic interference effects in operational amplifiers. *Electron. Lett.* **2007**, *43*, 84–85. [CrossRef]
12. Fiori, F.; Crovetti, P.S. Nonlinear effects of radio-frequency interference in operational amplifiers. *IEEE Trans. Circuits Syst. I Fundam. Theory Appl.* **2002**, *49*, 367–372. [CrossRef]
13. Yu, J.; Amer, A.; Sanchez-Sinencio, E. Electromagnetic Interference Resisting Operational Amplifier. *IEEE Trans. Circuits Syst. I Regul. Pap.* **2014**, *61*, 1917–1927. [CrossRef]
14. Hedayati, R.; Lanni, L.; Rodriguez, S.; Malm, B.G.; Rusu, A.; Zetterling, C.M. A Monolithic, 500 °C Operational Amplifier in 4H-SiC Bipolar Technology. *IEEE Electron Device Lett.* **2014**, *35*, 693–695.
15. Lanni, L.; Ghandi, R.; Malm, B.G.; Zetterling, C.M.; Ostling, M. Design and Characterization of High-Temperature ECL-Based Bipolar Integrated Circuits in 4H-SiC. *IEEE Trans. Electron Devices* **2012**, *59*, 1076–1083. [CrossRef]
16. Graffi, S.; Masetti, G.; Piovaccari, A. Criteria to reduce failures induced from conveyed electromagnetic interferences on CMOS operational amplifiers. *Microelectron. Reliab.* **1997**, *37*, 95–113. [CrossRef]
17. Richelli, A. CMOS OpAmp Resisting to Large Electromagnetic Interferences. *IEEE Trans. Electromagn. Compat.* **2010**, *52*, 1062–1065. [CrossRef]
18. Anjan Kumar Pudi, N.S.; Redouté, J.M.; Baghini, M.S. A Generic EMI-Immune Technique for Differential Amplifiers with Single-Ended Output. *IEEE Trans. Electromagn. Compat.* **2018**, *60*, 958–964. [CrossRef]
19. Zhou, Z.; Zhu, L.; Yang, R.; Li, J.; Wang, W.; Pan, J.; Liu, M.; Wang, K.; Wang, Z. A High CMRR Instrumentation Amplifier Employing Pseudo-Differential Inverter for Neural Signal Sensing. *IEEE Sens. J.* **2022**, *22*, 419–427. [CrossRef]
20. Richelli, A.; Colalongo, L.; Kovacs-Vajna, Z.M. Increasing the immunity to electromagnetic interferences of CMOS OpAmps. *IEEE Trans. Reliab.* **2003**, *52*, 349–353. [CrossRef]
21. Li, W.; Wei, G.; Pan, X.; Wan, H.; Lu, X. Electromagnetic Compatibility Prediction Method Under the Multifrequency in-Band Interference Environment. *IEEE Trans. Electromagn. Compat.* **2018**, *60*, 520–528. [CrossRef]
22. Su, D.; Cai, S.; Li, Y.; Li, W.; Diao, X. A Novel Equivalent Construction Method Applying to Radar EMS Threshold Test Under Multi-Source Heterogeneous Electromagnetic Environment. *IEEE Trans. Electromagn. Compat.* **2021**, *63*, 1910–1920. [CrossRef]
23. Tang, S.; Zhou, Z. Evaluation of Electromagnetic Environmental Effects for Ship System Based on Impact Degree. *IEEE Lett. Electromagn. Compat. Pract. Appl.* **2020**, *2*, 129–133. [CrossRef]
24. Farina, L.; Mestriner, D.; Procopio, R.; Brignone, M.; Delfino, F. The Lightning Power Electromagnetic Simulator for Transient Overvoltages (LIGHT-PESTO) Code: A User-Friendly Interface With the MATLAB-Simulink Environment. *IEEE Lett. Electromagn. Compat. Pract. Appl.* **2020**, *2*, 119–123. [CrossRef]
25. Wang, J.; Fujiwara, O.; Uda, T. New approach to safety evaluation of human exposure to stochastically-varying electromagnetic fields. *IEEE Trans. Electromagn. Compat.* **2005**, *47*, 971–976. [CrossRef]
26. Hubing, T.H.; Drewniak, J.L.; Doren, T.P.V.; Hockanson, D.M. Power bus decoupling on multilayer printed circuit boards. *IEEE Electromagn. Compat. Mag.* **2020**, *9*, 66–77. [CrossRef]
27. Kim, J.; Lee, J.; Ahn, S.; Fan, J. Closed-Form Expressions for the Noise Voltage Caused by a Burst Train of IC Switching Currents on a Power Distribution Network. *IEEE Trans. Electromagn. Compat.* **2014**, *56*, 1585–1597. [CrossRef]
28. Dhia, S.B.; Ramdani, M.; Sicard, E. *Electromagnetic Compatibility of Integrated Circuits*; Springer: New York, NY, USA, 2006; pp. XIII, 473.
29. *MIL-STD-461G*; Requirements for the Control of Electromagnetic Interference Characteristics of Subsystems and Equipment. Department of Defense Interface Standard: Arlington, VA, USA, 2015.
30. *GJB 151B-2013*; Electromagnetic Emission and Susceptibility Requirements and Measurements for Military and Subsystems. Electronic Information Foundation Department of the General Equipment Department of the Chinese People's Liberation Army: Beijing, China, 2013.
31. Redouté, J.M.; Steyaert, M. *EMC of Analog Integrated Circuits*; Springer: Dordrecht, The Netherlands, 2010; pp. 142–182.
32. *IEC 62132-4*; Integrated Circuits—Measurement of Electromagnetic Immunity—Part 4: Direct RF Power Injection Method. International Electrotechnical Commission: Geneva, Switzerland, 2006.
33. *IEC 62132-1*; Integrated Circuits, Measurement of Electromagnetic Immunity, 150 KHz–1 GHz: General Conditions and Definitions—Part 1. International Electrotechnical Commission Standard: Geneva, Switzerland, 2007.

micromachines

Article

Compact Physics Hot-Carrier Degradation Model Valid over a Wide Bias Range

Stanislav Tyaginov [1,*], Erik Bury [1], Alexander Grill [1], Zhuoqing Yu [2], Alexander Makarov [1], An De Keersgieter [1], Mikhail Vexler [3], Michiel Vandemaele [1], Runsheng Wang [2], Alessio Spessot [1], Adrian Chasin [1] and Ben Kaczer [1]

1 Imec, Kapeldreef 75, 3001 Leuven, Belgium; erik.bury@imec.be (E.B.); alexander.grill@imec.be (A.G.); alexander.makarov@imec.be (A.M.); an.dekeersgieter@imec.be (A.D.K.); alessio.spessot@imec.be (A.S.); adrian.chasin@imec.be (A.C.); ben.kaczer@imec.be (B.K.)
2 Institute of Microelectronics, Peking University, Beijing 100871, China; zhuoqing_yu@163.com (Z.Y.); r.wang@pku.edu.cn (R.W.)
3 A.F. Ioffe Institute, Polytechnicheskaya 26, 194021 St.-Petersburg, Russia; vexler@mail.ioffe.ru
* Correspondence: stanislav.tyaginov@imec.be

Abstract: We develop a compact physics model for hot-carrier degradation (HCD) that is valid over a wide range of gate and drain voltages (V_{gs} and V_{ds}, respectively). Special attention is paid to the contribution of secondary carriers (generated by impact ionization) to HCD, which was shown to be significant under stress conditions with low V_{gs} and relatively high V_{ds}. Implementation of this contribution is based on refined modeling of carrier transport for both primary and secondary carriers. To validate the model, we employ foundry-quality n-channel transistors and a broad range of stress voltages $\{V_{gs}, V_{ds}\}$.

Keywords: hot-carrier degradation; compact physics model; secondary carriers; impact ionization; interface traps; carrier transport

Citation: Tyaginov, S.; Bury, E.; Grill, A.; Yu, Z.; Makarov, A.; De Keersgieter, A.; Vexler, M.; Vandemaele, M.; Wang, R.; Spessot, A.; Chasin, A.; Kaczer, B. Compact Physics Hot-Carrier Degradation Model Valid over a Wide Bias Range. *Micromachines* **2023**, *14*, 2018. https://doi.org/10.3390/mi14112018

Academic Editor: Zhiming Chen

Received: 6 October 2023
Revised: 23 October 2023
Accepted: 25 October 2023
Published: 30 October 2023

1. Introduction

Hot-carrier degradation (HCD) has been recognized as the most harmful degradation issue limiting the lifetime of modern metal-oxide semiconductor field-effect transistors (MOSFETs) [1–3]. As such, comprehensive and predictive modeling of HCD is crucial for enabling further development of micro/nanoelectronics. Due to the complexity of the physical mechanisms underlying HCD [4,5], available physics-based models for HCD are computationally expensive [6–11]. On the other hand, empirical and phenomenological models [12–21] lack predictive capabilities because they do not capture the entire physical picture behind HCD. Hence, they cannot ensure that device lifetime under the operating regime is adequately predicted based on available experimental data acquired under more aggressive stress conditions, with most probably another dominant mechanism driving HCD.

In order to reach a compromise between model accuracy and optimized computational resources, we recently developed a compact physics model (CPM) for HCD [22], which was implemented on top of the reliability simulator Comphy [23]. This CPM relied on a simplified description of carrier transport for primary carriers and was shown to capture HCD in short-channel FETs stressed under the worst-case conditions (WCC) of HCD, i.e., under $V_{gs} = V_{ds}$ (V_{gs} and V_{ds} are gate and drain voltages, respectively) [24–27]. However, the model considered only the contribution of primary carriers to HCD, and the contribution of secondary carriers generated by impact ionization (II) was not implemented. On the other hand, in our recent publications, we demonstrated both experimentally [28] and theoretically [29,30] that under HC conditions with V_{gs} substantially lower than V_{ds}, secondary carriers provide a strong contribution to HCD. Moreover, secondary carriers generated by impact ionization were shown to give rise to the so-called turn-around effect when contributions to total damage related to primary and secondary carriers partially compensate each other. For example, screening of damage produced by primary electrons by the degradation component driven by secondary holes in lightly-doped drain nMOSFETs was reported by Vuillaume et al. [31]. Next, Chen et al. [32] demonstrated the same phenomenon in high-voltage drain-extended metal-oxide-semiconductor transistors subjected to hot-carrier degradation. Furthermore, Starkov et al. [33] performed an analysis of the turn-around effect in planar 5 V nMOSFETs based on results obtained with the charge-pumping technique; in a more recent paper [34], they carried out modeling of this phenomenon. Such an intricate behavior (when primary and secondary carriers generate/populate traps located in different device sections) was shown by various groups to be typical also for OFF-state stress [35–37]. Finally, in our recent work, we demonstrated that II can be the reason for the stimulated recovery of bias temperature instability (BTI) induced by HCD [38]. Therefore, the main goal of this work is to extend our CPM for HCD by incorporating the impact of secondary carriers on HCD.

Another important improvement of the CPM presented in this paper is the refinement of carrier transport treatment. Indeed, in the previous version of our CPM, average carrier energy was evaluated via the homogeneous energy balance equation used in drift-diffusion (DD) models [39], i.e., this energy was determined by the square of the electric field and the carrier mobility. However, the DD approach to the Boltzmann transport equation (BTE) solution is known to fail to model carrier transport in ultra-scaled FETs [40,41]. Therefore, implementation of the contribution of secondary carriers should rely on refined carrier transport treatment for both primary and secondary carriers [42]. The extended CPM is validated here against HCD data over a broad $\{V_{gs}, V_{ds}\}$ range.

2. Experimental

To validate the model, we used planar n-channel MOSFETs (with primary and secondary carriers being electrons and holes, respectively). Note that we intentionally used planar MOSFETs with simplified geometry. The reason behind this is that transistors of novel architectures—such as fin [43,44], nanowire [45,46], nanosheet [47,48], forksheet [49–51], and complementary FETs [52,53]—have confined channels. As a consequence, modeling HCD in such devices would result in additional challenges due to quantum confinement effects and the 3D nature of the FET structure. In this study, however, we focus on the CPM for HCD and strive to minimize the complexity originating from "side effects". Employed transistors are foundry-quality devices with a channel length of L_g = 28 nm and an operating voltage of V_{dd} = 1.2 V. Their high-k gate stack is made of silica and hafnia layers with an equivalent oxide thickness of 1.3 nm. The devices were stressed under the worst-case conditions (WCC) for HCD in short-channel MOSFETs, i.e., at V_{gs} equal to V_{ds}; for both voltages, we used values of 1.8, 1.9, and 2.0 V. We also subjected these MOSFETs to HC stress at much lower V_{gs}, namely at V_{gs} of 1.0 V (V_{ds} was chosen to be equal to 1.9, 2.0, and 2.1 V) and V_{gs} = 0.69 V (V_{ds} values were equal to 1.8, 1.9, and 2.0 V). All experiments were conducted at room temperature with stress times of up to 144 s.

To assess HCD, we monitored relative changes ($\Delta I_{d,lin}$) of the drain current in the linear regime (with V_{ds} = 50 mV and V_{gs} = 1.2 V) as a function of stress time (t). Recorded $\Delta I_{d,lin}$ values were relative, i.e., normalized to the drain current in the pristine MOSFET. To enable fast measurements of many samples in parallel we used on-chip smart arrays [54,55]. Consequently, for each combination of V_{gs} and V_{ds} we employed ~3800 samples, obtained $\Delta I_{d,lin}$ changes, and then for each stress time step, we extracted their mean values. Further in the paper under $\Delta I_{d,lin}$, we understand these mean values, which are summarized in Figures 1–3. Our compact physics model was verified in order to reproduce these mean $\Delta I_{d,lin}(t)$ traces.

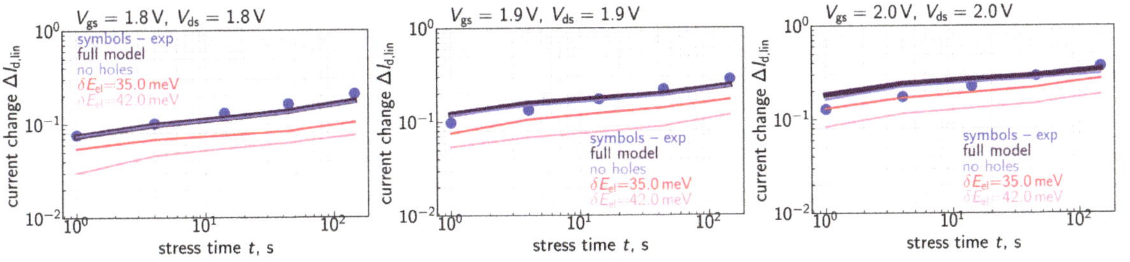

Figure 1. Experimental and calculated $\Delta I_{\mathrm{d,lin}}(t)$ traces for the WCC of HCD with V_{gs} = V_{ds} = 1.8, 1.9, and 2.0 V. To analyze the role of secondary holes also $\Delta I_{\mathrm{d,lin}}(t)$ traces obtained without their contribution were evaluated. We use the energy loss parameter of δE = 28 meV, but the model results are very sensitive to its variations (see Figures 10 and 13 for the impact of the energy loss parameter) and therefore $\Delta I_{\mathrm{d,lin}}(t)$ curves for δE = 35 and 43 meV are also shown.

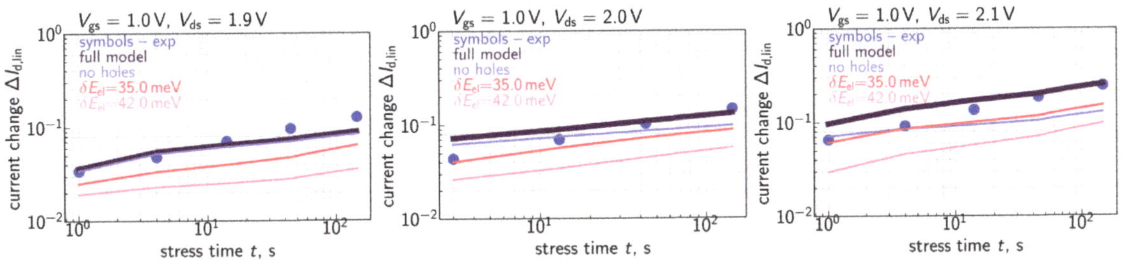

Figure 2. The same as in Figure 1 but for V_{gs} = 1.0 V and V_{ds} = 1.9, 2.0, and 2.1 V.

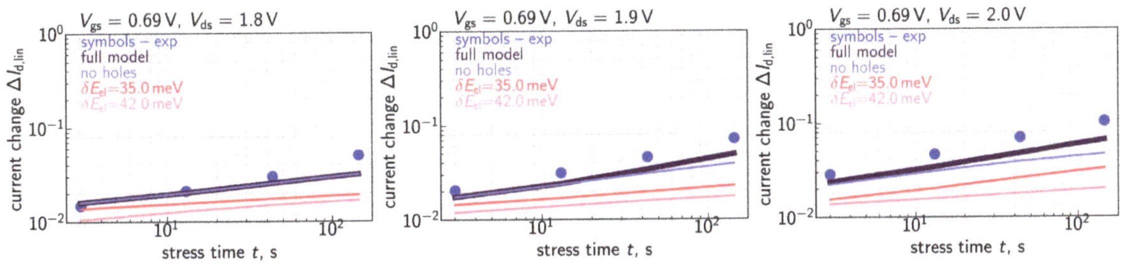

Figure 3. The same as in Figure 1 but for V_{gs} = 0.69 V and V_{ds} = 1.8, 1.9, and 2.0 V.

Note that the extraction of $\Delta I_{\mathrm{d,lin}}$ drifts was based on measurements of entire $I_{\mathrm{d}} - V_{\mathrm{gs}}$ curves with the gate voltage sweeping from 0 to 1.2 V (meanwhile, the stress phase was interrupted). Such a procedure required a measurement time of ~0.75 s, and therefore HCD was assessed with the corresponding delay. Although recovery of HCD (or more precisely, passivation of P_b centers that were created by the rupture of Si-H bonds [56–59]) was reported by several groups [60–64], this process was shown to have a significant rate only at temperatures of 150 °C or higher. As our experiments were conducted at room temperature, we can conclude that $\Delta I_{\mathrm{d,lin}}$ values did not recover during the aforementioned measurement delay. Another recoverable contribution to the entire damage can originate from the trapping of carriers by defects in the dielectric layer (bias temperature instability) [65,66]. However, this type of degradation is known to be homogeneously distributed over the coordinates along the Si/SiO_2 interface. Our recent experimental studies have shown that under the same stress conditions, as in this paper applied to the same devices, the factor of degradation localization is within the range of 0.6–0.8 (a value of this factor equal to 1 corresponds to strong damage localization near the drain). In other words, this type of damage is relatively strongly localized, and therefore the contribution of bias temperature instability can be neglected [55]. The fact that this localization factor is less than 1 stems from the contribution to HCD provided by secondary holes with the position of the corresponding interface trap density maximum shifted towards the source, as compared with the near-drain N_{it} maximum related to primary electrons (see Section 4). Moreover, our experience in the field of BTI suggests that significant BTI recovery occurs after relaxation time, which is an order of magnitude longer than stress time [67]. This is not the case for our measurements because relaxation time was ~0.75 s, but the shortest stress duration was 1 s. Hence, even though BTI provides a non-negligible contribution to the total damage, one can neglect the recovery of $\Delta I_{d,lin}$ values.

In our study, we used very high stress voltages, V_{gs} and V_{ds}. There are several reasons for this. According to our understanding, the physical mechanism behind HCD is the dissociation of Si-H bonds at the Si/SiO_2 interface induced by channel carriers [6,27,68]. The bond dissociation reaction has two pathways, i.e., the single- and multiple-carrier (SC and MC, respectively) mechanisms of bond dissociation [4,69]. Although the MC-process is considered dominant for HCD under low-voltage stress conditions, it was shown that this mechanism can lead to a significant contribution to HCD even in high-voltage transistors [7,70,71]. The MC-process is driven by colder carriers, and a high carrier concentration typically results in a high rate of this mechanism. Therefore, we intentionally used high V_{gs} values to ensure that this process has a significant rate. As for the SC-process, it is driven by hot carriers whose energies are determined by the applied V_{ds}. To ensure that the SC-mechanism has a high rate as well, we applied large source-drain voltages. One of the goals of this study was to analyze the contribution of secondary carriers generated by II to HCD, and therefore high V_{ds} values were chosen to enhance this contribution. Finally, in our experiments, stress times were limited by 144 s, and therefore these $\{V_{ds}$ and $V_{gs}\}$ values were supposed to result in significant $\Delta I_{d,lin}$ changes within the aforementioned stress time window.

Let us mention that the time exponents featured by the measured $\Delta I_{d,lin}(t)$ curves (Figures 1–3) are within the range of 0.2–0.35 and therefore smaller than those reported for HCD by several other groups [72,73]. This is because in our study we used quite aggressive HC stress: one can see that under the lowest stress voltages of V_{gs} = 0.69 V and V_{ds} = 1.8 V (see Figure 2), already at a stress time of ~1 s the $\Delta I_{d,lin}$ value is ~2%, while under $V_{gs} = V_{ds}$ = 2.0 V (Figure 1), the drain current change $\Delta I_{d,lin}$ for t~1 s substantially exceeds 10%. We analyzed the behavior of HCD under high stress voltages in one of our previous papers [74]. It was shown that even at short stress times, the drain area of the transistor is already heavily degraded and the concentration of N_{it} is saturated, i.e., the available Si-H bonds are predominantly broken, and the near-drain N_{it} value does not vary with the coordinate along the interface. In this scenario, the further increase in $\Delta I_{d,lin}$ with t is due to propagation of the N_{it} front inside the device channel. As a result, the time exponent of the $\Delta I_{d,lin}(t)$ curves is somewhat lower than that typical for milder HCD. More severe stress conditions with a higher V_{gs} value result in a broader degraded region with almost constant N_{it} near the device drain, a more saturated HCD, and consequently a smaller time exponent. Such a behavior is consistent with HCD data published by Varghese et al. [75] and Yamagata et al. [76], where the authors have shown that the time slope of degradation traces reduces at higher stress voltages; the same trend was obtained within our TCAD model applied to HCD modeling in finFETs [77]. To conclude, we intentionally used very aggressive stress conditions and therefore our obtained $\Delta I_{d,lin}(t)$ dependencies were more gradual compared with those typically monitored during HCD.

3. The Model

Our CPM is based on the detailed physical picture underlying HCD, which was captured in the TCAD version of our HCD model [9,78]. Both versions of the model consider the dissociation of Si-H bonds at the Si/SiO_2 interface as the microscopic mechanism responsible for HCD. Such a rupture reaction can be driven by a solitary highly energetical carrier; this process is referred to as the single-carrier (SC) mechanism of bond breakage [68,69]. In other words, the model is consistent with the energy-driven paradigm described by Rauch, La Rosa, and Guarin [19,21]. Alternatively, a series of colder carriers can induce the multivibrational excitation of the bond, which results in its weakening and finally rupture; this scenario is referred to as the multiple-carrier (MC) mechanism of bond rupture, as proposed by the group of Hess [4,69,79,80]. In the case of ultra-scaled MOSFETs, it was shown that the most probable pathway of bond dissociation is via coupled MC- and SC-processes [8,9]. Therefore, to calculate the rates of the SC- and MC-mechanisms, one needs to solve the carrier transport sub-task of the entire problem of HCD modeling and obtain the energy distribution function (DF) for carriers. Throughout the paper, we consider HCD in an nMOSFET, which is sketched in Figure 4.

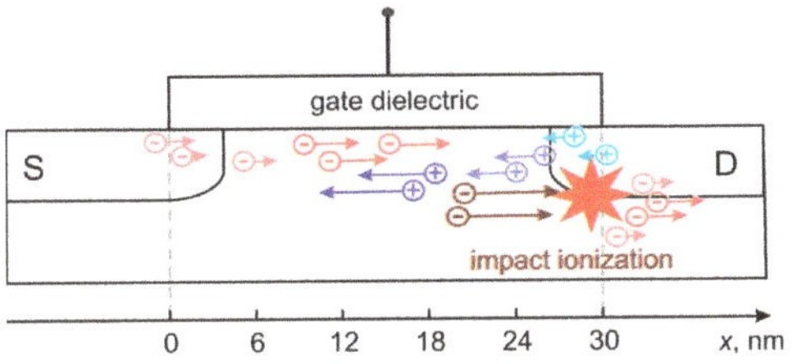

Figure 4. Schematic representation of the device used to validate the model. The device source is at $x = 0$. Primary carriers, which drive impact ionization, are electrons. Secondary holes generated by impact ionization near the drain are accelerated towards the source where they reach highest energies and provide the most significant contribution to HCD.

3.1. Transport of Primary Carriers

In the TCAD version of our HCD model [9], carrier DFs were obtained by solving the carrier BTE using the deterministic solver ViennaSHE [81–84], which solved BTE using the expansion of the carrier energy distribution function into a series of spherical harmonics [85–87]. Such a solution is computationally expensive, and therefore, in the CPM for both types of carriers, we use the analytical expression for the carrier DF $f(E)$ (with E being carrier energy), as proposed by Grasser et al. [88]:

$$f(E) = A \exp\left[-\left(\frac{E}{E_{\mathrm{ref}}}\right)^{b}\right] + C \exp\left[-\frac{E}{k_{\mathrm{B}} T_{\mathrm{L}}}\right], \tag{1}$$

where the first term represents the fraction of non-equilibrium (hot) carriers and the second term corresponds to the Maxwellian distribution of thermalized (cold) carriers; E_{ref} is the reference energy for hot carriers; A and C are weighting factors; k_{B} the Boltzmann constant; and T_{L} is the lattice temperature. The exponent b is chosen to be 1 within the source and drain regions and 2 elsewhere.

The transistor is represented by a series of slices in the source-drain direction. For each of the slices, we need to obtain the three adjustable parameters E_{ref}, A, and C, which determine the carrier DF [22]. For primary carriers, in each slice labeled with an index i, we solve the system of three equations that are based on three moments of the BTE, with the closures being the carrier concentration (n), average carrier energy (E_{e}), and the kurtosis (β_{e}):

$$\int_0^{\infty} f_{\mathrm{e},i}(E) g_{\mathrm{c}}(E) dE = n_i \tag{2}$$

$$\frac{1}{n_i} \int_0^{\infty} E f_{\mathrm{e},i}(E) g_{\mathrm{c}}(E) dE = E_{\mathrm{e},i} \tag{3}$$

$$\frac{3}{5} \frac{\langle E^2 \rangle}{\langle E \rangle^2} = \frac{3}{5} \frac{n_i \int_0^{\infty} E^2 f_{\mathrm{e},i}(E) g_{\mathrm{c}}(E) dE}{\left(\int_0^{\infty} E f_{\mathrm{e},i}(E) g_{\mathrm{c}}(E) dE\right)^2} = \beta_{\mathrm{e},i}. \tag{4}$$

For the density-of-states in the Si conduction band $g_c(E)$ in (2) and (3), we use the analytical expression based on the Kane non-parabolic dispersion relation [88].

The carrier concentration is computed using the Poisson solver implemented in the reliability simulator, Comphy. To calculate the carrier kurtosis, we use the empirical expression proposed by Grasser et al. [88], derived from the rigorous BTE solution (for more detail, see [22]). Within the previous version of our CPM [22], we calculated average electron energy (E_e) as

$$E_e = \frac{3}{2} k_B T_L + q \tau_{e,E} \mu_e F_{Si}^2, \quad (5)$$

where $\tau_{e,E}$ is the electron energy relaxation time, μ_e the electron mobility, F_{Si} the electric field in the channel, and q the elementary charge.

Equation (5) corresponds to the manner of carrier energy evaluation within the DD approach to the BTE solution, which leads to spurious results for short-channel FETs. Figure 5 shows E_e plotted as a function of the coordinate x along the Si/SiO_2 interface evaluated using (5) and the $E_e(x)$ profile obtained from the rigorous BTE solution with ViennaSHE. These $E_e(x)$ dependencies were obtained for an nMOSFET with L_g = 28 nm (the source is at x = 0 nm); see Section 2; the applied voltages are V_{gs} = V_{ds} = 1.8 V. One can see that these two profiles have different shapes. Even more, at a moderate V_{gs} of 1.8 V, the *average* electron energy obtained within the DD-based approach reaches a value of $\sim$10 eV, which is unphysical. Therefore, in the refined CPM for HCD, the carrier transport description needs to be revised.

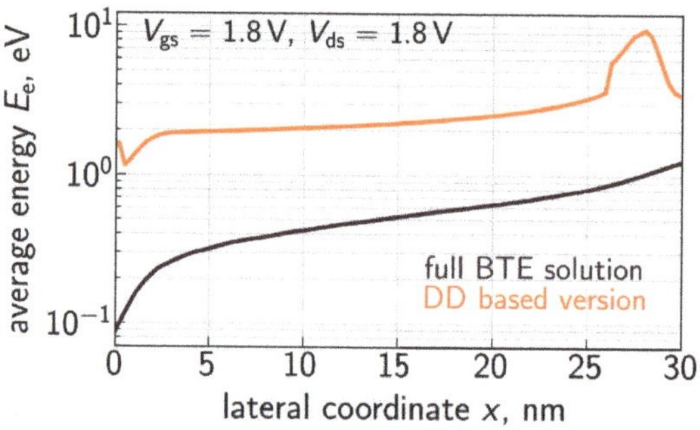

Figure 5. Average energy of primary electrons E_e as a function of the lateral coordinate x along the Si/SiO_2 interface of an n-channel MOSFET with a gate length of 28 nm (x = 0 nm corresponds to the source). Shown are two $E_e(x)$ profiles: one obtained with the DD-based approach and another one calculated employing the BTE solution with the carrier transport simulator ViennaSHE. This comparison illustrates that the DD based approach to estimation of average carrier energy, and further to HCD modeling, is not applicable.

Within the refined transport modeling approach (Figure 6), we consider two competing mechanisms, i.e., carrier acceleration by the electric field and energy loss due to scattering. Energy gained by carriers is determined by the band bending profile in the source-drain direction, and this profile is obtained from the Poisson solver of Comphy. For compact physics treatment of scattering mechanisms, we assume that a carrier loses an amount of energy δE (=28 meV) each time it passes a distance equal to its mean free path λ. In the simplified version of the model, we use $\lambda = 3$ nm; this value is consistent with the electron mean free path reported in [89].

Let us comment on the choice of the parameters λ_0 and δE. The carrier mean free path is determined by the scattering rate and the carrier velocity. These two quantities are a function of energy, and therefore, for a thorough evaluation of the mean free path, we need to consider the actual energy DF of the carrier ensemble. Based on our experience in modeling carrier transport in sub-100 nm MOSFETs, we envisage that DFs of substantially hot carriers feature a plateau (i.e., in this local energy range, the DF is a weak function of carrier energy) spreading up to energies of $\sim|qV_{ds}|$ [78,90]. HCD measurements in modern scaled MOSFETs are conducted at V_{ds} lying in the range of $[1.0, 2.0]$ V. In this energy range, the electron-phonon scattering rate is $\sim$(0.5–1.0) $\times$ $10^{14}\,s^{-1}$ [91–93]. The hole-phonon scattering rates in the valence band also have comparable values [93]. As for scattering at ionized impurities, Qiu et al. showed that this mechanism is dominant at carrier energies not exceeding $\sim$0.3 eV, while at higher energies, electron-phonon interactions prevail [94]. Therefore, for our estimation, ionized impurity scattering can be neglected. The hole and electron velocities in the aforementioned energy segment are in the range of $\sim$(0.5–1.0) $\times$ 10^8 cm/s [91,95]. By combining the given scattering rates and carrier velocities, we obtain the carrier mean free path to be within an interval of 2–10 nm, i.e., the value $\lambda = 3$ nm used in our CPM is consistent with our estimation.

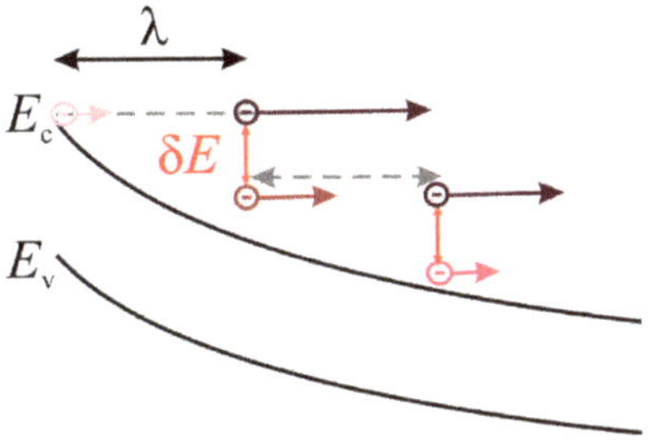

Figure 6. In the refined CPM for HCD, the average energy of electrons is determined by carrier acceleration by the electric field and energy dissipation due to scattering. Energy gained from the electric field is evaluated based on the band bending profile in the transport direction. We assume that when an electron travels a distance equal to its mean free path (λ) it loses a certain amount of energy δE.

The values of energy loss due to the interaction of an electron with an optical phonon are 62.0 and 58.6 meV for longitudinal and transverse optical modes, respectively [95]. For acoustic phonons, these values are 12.1 and 19.0 meV for the two branches of transverse acoustic phonons and 18.4 and 47.4 meV for the longitudinal acoustic phonon branches [95]. Based on the listed values, scattering at acoustic phonons is often considered quasi-elastic, and the corresponding contribution to carrier energy loss is hence neglected. However, recently, Fischetti et al. [92] suggested that this assumption should be revised because, although energy loss due to optical phonon scattering is higher than that typical for scattering at acoustic phonons, the rate of the former mechanism is significantly lower than in the latter case. This idea is consistent with previously published data [96]. Therefore, our energy loss parameter $\delta E = 28$ meV is a reasonable trade-off between energy loss values typical for optical and acoustic phonons.

Figure 7 compares the $E_e(x)$ profiles calculated for V_{gs} = 1.0 V and V_{ds} = 2.1 V using the refined carrier transport model and the DD-based approach of Equation (5). One can see that the former profile is quantitatively similar to that obtained using ViennaSHE (Figure 5) and the maximum E_e is $\sim$1.7 eV, i.e., reasonable for V_{ds} = 2.1 V. Quite to the contrary, the profile evaluated with the DD-based approach reaches an energy of more than 20 eV, thereby manifesting the inapplicability of Formula (5).

With E_e calculated for each transistor slice, we solve the system (2)–(4) and obtain the electron energy DF f_e. An example of generalized electron DFs (i.e., $f_{e,i}(E)g_c(E)$ with dimensionality of $J^{-1}m^{-3}$) for V_{gs} = 1.0 V and V_{ds} = 2.1 V is shown in Figure 8 for five different positions at the Si/SiO_2 interface with x = 0.2, 10.2, 20.2, 25.0, and 26.5 nm. The position x = 0.2 nm corresponds to the source area, where electrons are thermalized, and therefore the DF is Maxwellian. As x changes towards the drain, the DFs shift from equilibrium, which is manifested by the extension of the plateau (with DF values being almost unchanged with increasing E). For example, at x = 26.5 nm, when almost the entire gate voltage V_{ds} drops across the channel, this plateau propagates up to $\sim$1.9 eV. This DF transformation is consistent with the behavior of electron DFs obtained using ViennaSHE; see [9].

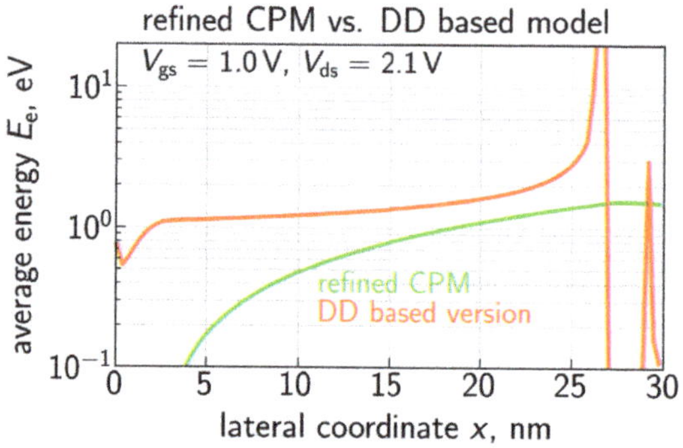

Figure 7. Average energy of primary electrons vs. the lateral coordinate x obtained with the refined CPM and compared with that evaluated using the DD based approach.

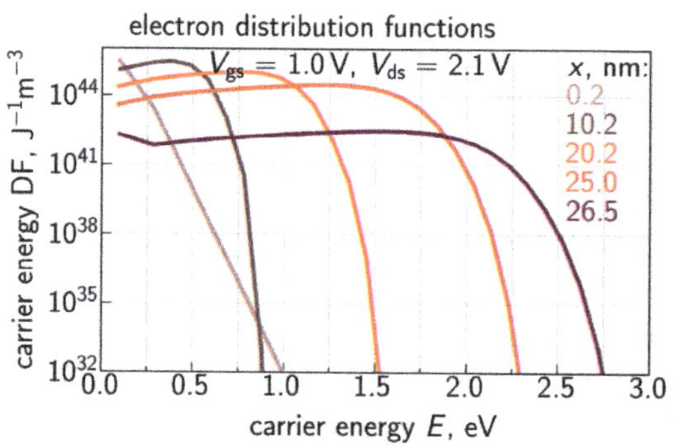

Figure 8. Generalized distribution functions of primary electrons obtained with the refined CPM for different positions along the Si/SiO_2 interface.

3.2. *Transport of Secondary Carriers*

With the obtained DFs for primary electrons, we proceed to the modeling of carrier transport for secondary holes. Secondary carriers are generated by impact ionization, and to evaluate the II rate (G_{II}), we use the model by Grasser et al. [88]:

$$G_{II} = \int P_{II}(E) f_e(E) g_c(E) \mathrm{d}E, \tag{6}$$

where the reaction rate $P_{II}(E)$ is

$$P_{II}(E) = P_0 \left(\frac{E - E_{th}}{E_{th}} \right)^2 \tag{7}$$

with $E_{th} = 1.12$ eV, i.e., equal to the band gap of Si and $P_0 = 4.18 \times 10^{12}\,s^{-1}$.

A comparison of the G_{II} dependencies calculated with the refined CPM and the DD-based approach is given in Figure 9. Whereas the former profile is in good qualitative agreement with the G_{II} results from [88], the latter one substantially deviates from them in terms of the $G_{II}(x)$ shape and peak values. Therefore, using the G_{II} rate obtained with the DD based approach would result in a severely overestimated contribution of secondary holes and hence spurious $\Delta I_{d,lin}$ values. Figure 10 provides a summary of $G_{II}(x)$ profiles evaluated with different values of energy loss: δE = 28, 35, and 42 meV. The increasing value of δE results in a lower average energy of the electron ensemble and therefore a smaller rate G_{II}.

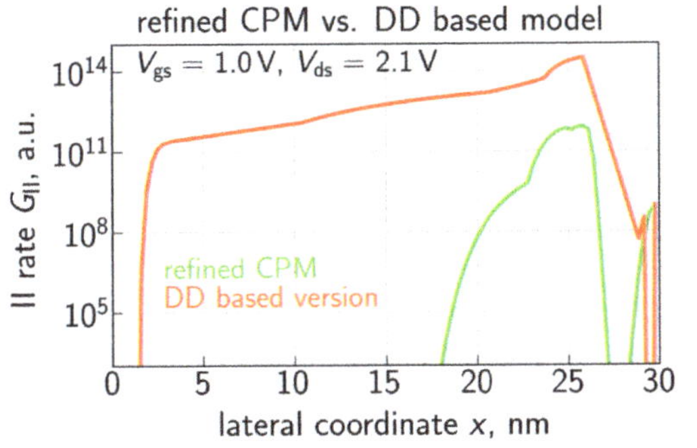

Figure 9. The impact ionization rate G_{II} as a function of x calculated with the refined CPM and the DD based formula (the latter approach overestimates G_{II}).

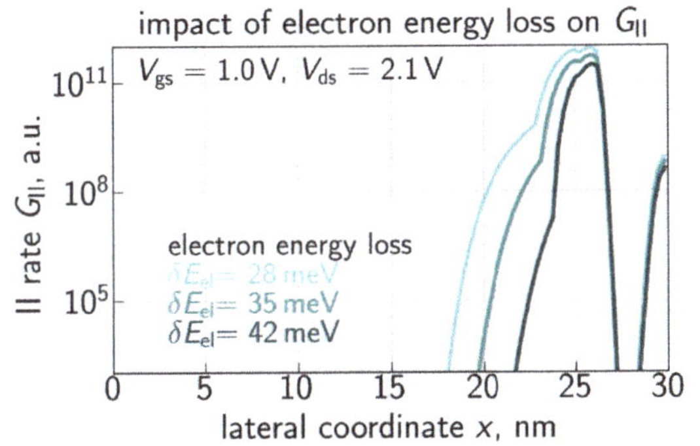

Figure 10. The II rate G_{II} vs. the coordinate x obtained for three different values of the energy loss parameter δE = 28, 35, and 42 meV.

For secondary carriers, which are generated by impact ionization, we do not have access to the hole concentration (p), and therefore, instead of the system (2)–(4), we employ a modified set of equations:

$$\frac{1}{p_i}\int_0^\infty E f_{h,i}(E) g_v(E) dE = E_{h,i} \tag{8}$$

$$\frac{3}{5}\frac{\langle E^2\rangle}{\langle E\rangle^2} = \frac{3}{5}\frac{p_i \int_0^\infty E^2 f_{h,i}(E) g_v(E) dE}{\left(\int_0^\infty E f_{h,i}(E) g_v(E) dE\right)^2} = \beta_{h,i} \tag{9}$$

$$J^{out}_{h,i} = J^{in}_{h,i-1} + G_{II,i} l_s - R_i l_s. \tag{10}$$

In this system, Equation (10) is the flux balance equation for holes. For each slice i, we assume that the supply of secondary holes should be equal to the loss of holes; see Figure 11. The supply components are hole generation by II, designated as $G_{II,i} l_s$ (l_s is the slice length), and the hole flux $J^{in}_{h,i-1}$ from the previous slice with index $i-1$. Note that in the case of holes, the slice enumeration begins at the drain (this slice has i = 0), and the index i increases towards the source (Figure 11). Hole loss is due to recombination with the rate R_i and the flux $J^{out}_{h,i}$ of holes departing from the slice i to the slice $i+1$. The flux of holes entering slice i (see Figure 11) is calculated as

$$J^{in}_{h,i-1} = \frac{2}{\pi}\int f_{h,i-1}(E) g_v(E) v_{h,i-1}(E) \mathrm{d}E, \tag{11}$$

where $f_{h,i}$ is the hole energy DF, g_v is the density of states in the valence band, and $v_{h,i}$ is the velocity of holes. The coefficient $2/\pi$ is related to the averaging of $\cos\theta$ (where θ is the angle between the carrier velocity and the transport direction) over a uniform distribution of $\theta \in [-\pi/2; \pi/2]$. Holes leaving the slice i (flux $J_{h,i}^{out}$ in Figure 11) can move in any direction and therefore the coefficient $2/\pi$ is omitted:

$$J_{h,i-1}^{out} = \int f_{h,i-1}(E) g_v(E) v_{h,i-1}(E) dE, \tag{12}$$

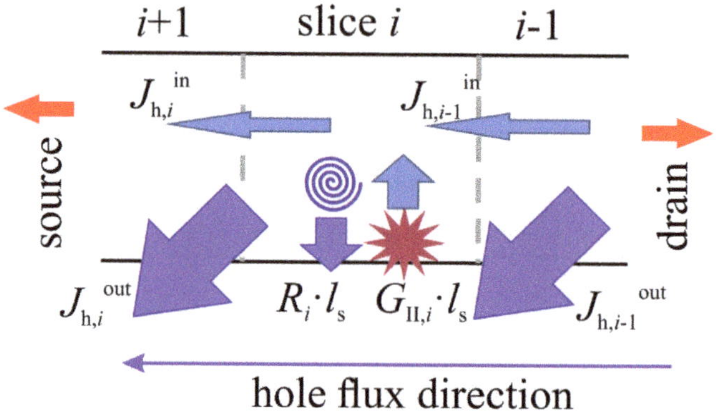

Figure 11. Schematic representation of the hole flux balance equation. For each transistor slice with an index i (for holes the slice enumeration begins at the drain, where the holes are predominantly created by II, i.e., i increases towards the source) a balance between hole supply and hole loss is considered. The hole supply component is due to holes arriving from the previous slice $i-1$ (the hole flux $J_{h,i-1}^{in}$) and impact ionization (with the rate G_{II}), while hole loss is due to hole departure to the slice $i+1$ ($J_{h,i}^{out}$) and recombination (with the rate R). The slice length is designated as l_s.

The hole concentration p enters (9) and it is evaluated as

$$p_i = \int f_{h,i}(E) g_v(E) dE. \tag{13}$$

For the recombination rate R_i we assume that the concentration of secondary holes is much less than that of primary electrons, i.e., $p \ll n$, and therefore [97]:

$$R_i = p_i / \tau_{E,h}, \tag{14}$$

where $\tau_{E,h}$ is the energy relaxation time for holes.

Hole DFs obtained by solving the system of Equations (8)–(10) for V_{gs} = 1.0 V and V_{ds} = 2.1 V are plotted in Figure 12 for different positions along the interface. One can see that for the drain area at x = 26 nm (this x value corresponds to the II rate peak, Figure 10), where holes are predominantly generated, the DF is Maxwellian because holes are thermalized. However, holes are accelerated by the electric field towards the source and their DFs become strongly non-equilibrium. The impact of the carrier energy loss δE on hole DFs is depicted in Figure 13 for two positions along the interface; like in Figure 10, we used δE = 28, 35, and 42 meV. One can see that the increasing δE leads to lower values of DFs in the entire energy range. This trend appears to be very reasonable because a higher δE results in a lower II rate (Figure 10), thereby decreasing the hole concentration, and holes themselves lose more energy, i.e., become colder. Finally, Figure 14 provides a comparison of hole DFs calculated with the refined carrier transport treatment and within the DD-based approach. It can be seen that the latter DFs have enormously high values, and their behavior is consistent with the spurious II rate (Figure 9) obtained using (5).

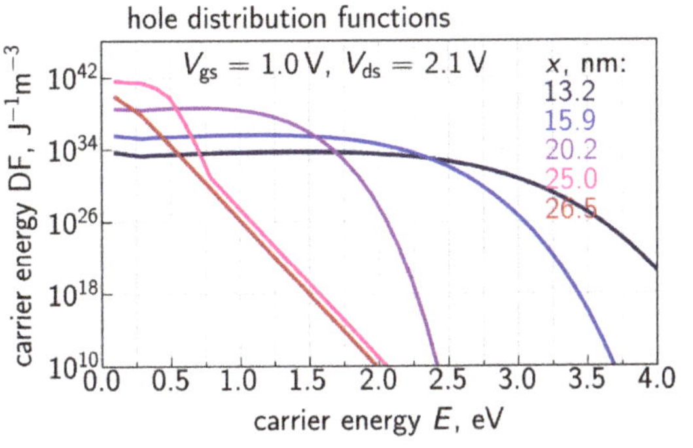

Figure 12. Generalized DFs for secondary holes calculated with the refined compact physics model and plotted for different positions at the Si/SiO_2 interface.

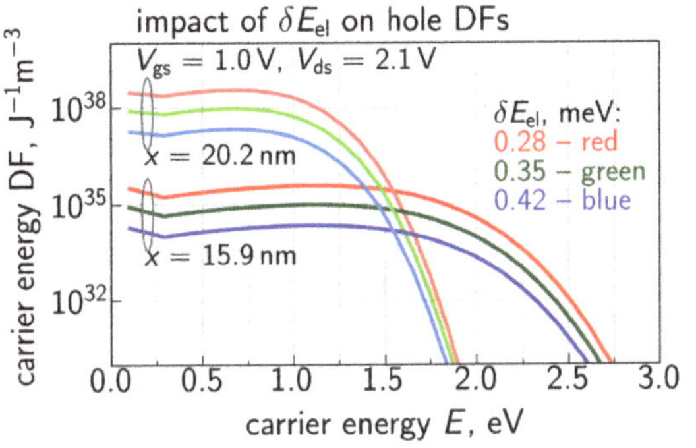

Figure 13. Impact of the energy loss parameter δE on the secondary hole DFs. At larger δE both types of carriers are colder and this trend is confirmed by the hole DFs.

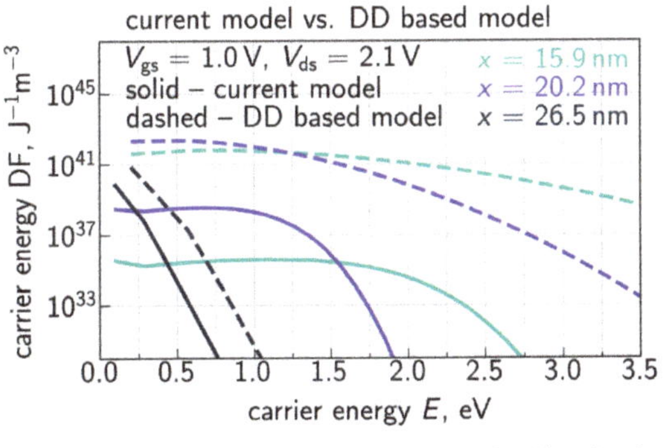

Figure 14. Comparison of hole DFs obtained with refined carrier transport treatment and using the DD based model. In the latter case the DFs have spuriously high values.

The evaluated DFs for both types of carriers are then used to calculate bond dissociation rates and the interface trap density N_{it} as a function of the lateral coordinate x for each stress time step t; for details, see [9,78]. The $N_{it}(x)$ profile is then employed to calculate $\Delta I_{d,lin}(t)$ traces taking into account both electrostatic perturbation of the stressed device and mobility reduction; this procedure is described in [22].

It is important to emphasize that the developed CPM for HCD allows one to dramatically reduce computational time. The most computationally expensive part of our TCAD model for HCD is transport simulation. Depending on the device architecture (its complexity, the number of mesh points, etc.) and stress conditions, solving the BTE for a real device structure may require a few hours. For example, transport simulations carried out for the MOSFET employed in this study took approximately 3 h on a desktop. As for the CPM, its accuracy depends on the number of slices used to reproduce the transistor. However, increasing the number of slices would increase computational time (roughly) proportionally. Thus, it is important to find a balance between the number of slices (computational time) and model accuracy. For the calculations presented in this work, we used a relatively large number of slices, namely 100. As the gate length of our devices is 28 nm, using 100 slices results in a good resolution comparable to that provided by fine meshes used in commercial device simulators. In this case all calculations (including transport modeling, calculations of the N_{it} density, and finally obtaining $\Delta I_{d,lin}(t)$ traces) were completed within 1–2 min on a laptop.

4. Degradation Characteristics

For the case of the HCD WCC in short-channel transistors with $V_{gs} = V_{ds} = 1.8, 1.9$, and 2.0 V, $\Delta I_{d,lin}(t)$ traces are summarized in Figure 1. One can see that the refined model can capture experimental data with good accuracy. $\Delta I_{d,lin}(t)$ curves modeled with increased values of the energy loss parameter δE of 35 and 42 meV have lower values than those simulated with $\delta E = 28$ meV. This tendency is consistent with the impact of δE on the II rate G_{II} (Figure 10) and hole DFs (Figure 13). We also simulated $\Delta I_{d,lin}(t)$ dependencies disregarding the contribution of secondary holes, and one can see that these traces coincide with those obtained using the "full" model. In other words, if we neglect the HCD component driven by holes, we do not underestimate $\Delta I_{d,lin}$ changes, i.e., at $V_{gs} = V_{ds}$, the impact of secondary holes is not significant.

Quite to the contrary, at a much lower V_{gs} of 1.0 V, neglecting the contribution of secondary holes results in substantial underestimation of HCD (see Figure 2). The same behavior is also pronounced at $V_{gs} = 0.69$ V (Figure 3), but in a less prominent way. Such a trend can be understood considering that the secondary holes are generated by II, whose rate features a maximum at $V_{gs} = (0.4–0.5)V_{ds}$ [97]. This interrelation of the voltages corresponds to the stress conditions with $V_{gs} = 1.0$ V and, to a lesser extent, to the regimes with $V_{gs} = 0.69$ V being in mismatch with the WCC.

Another important peculiarity noticeable in Figures 2 and 3 is that the contribution of secondary holes becomes more significant at higher V_{ds}. This is because with an increasing V_{ds}, primary carriers become hotter, thereby resulting in a higher II rate and a higher concentration of secondary carriers; the secondary carriers also reach higher energies at higher V_{ds}. This trend is confirmed by $N_{it}(x)$ profiles obtained with and without the secondary hole contribution for $V_{gs} = 1.0$ V and $V_{ds} = 1.9$ and 2.1 V, see Figure 15. From Figure 15, we conclude that, in addition to the "traditional" N_{it} peak located at the drain and originating from primary carriers [98–100], secondary holes result in an N_{it} peak situated near the source, whose position is consistent with the results obtained using the TCAD model based on the rigorous BTE solution [30,101].

It is noteworthy that although the contribution to HCD provided by secondary holes is most prominent at the highest V_{ds}, it results in a change in the time slope of modeled $\Delta I_{d,lin}(t)$ traces. The data set used in this study was acquired within the time window limited by 144 s; however, at longer stress times (e.g., several ks), deviations of $\Delta I_{d,lin}$ values calculated disregarding the impact of secondary carriers from experimental $\Delta I_{d,lin}$ changes would be quite substantial. On the other hand, the major task the model aims at tackling is to—based on experimental data acquired at high stress voltages—predict device time-to-failure for regimes with operating voltages comparable to V_{dd}. Even though V_{dd} in the employed MOSFETs is 1.2 V, and therefore, in the aforementioned regimes, the impact of secondary carriers is weak, the model validation/calibration disregarding their contribution would result in a spurious value of device lifetime.

Let us emphasize that the model can thoroughly reproduce experimental $\Delta I_{\text{d,lin}}(t)$ traces for all stress conditions.

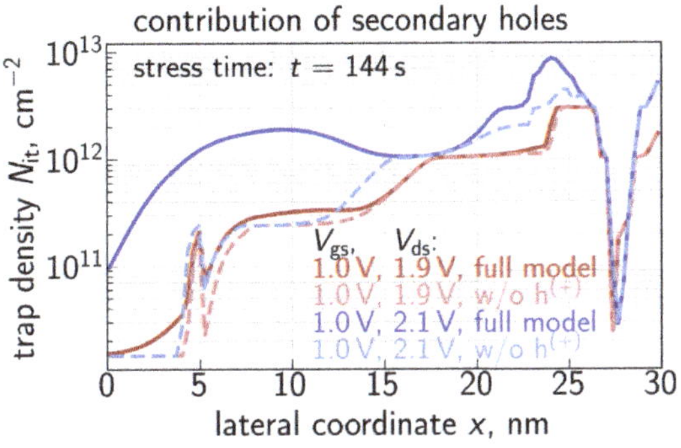

Figure 15. The interface trap density N_{it} as a function of the coordinate along the interface x for V_{gs} = 1.0 V and two values of V_{ds} = 1.9 and 2.1 V calculated with and without the contribution of secondary holes. One can see that the hole contribution results in the secondary N_{it} peak situated near the source, and this peak becomes more pronounced at higher V_{ds}.

5. Conclusions

We extended our compact physics model for hot-carrier degradation by implementing the component driven by secondary carriers generated by impact ionization. This implementation is based on refined carrier transport modeling for both types of carriers. Note that in the previous version of our CPM for HCD, the average carrier energy was estimated using the homogeneous energy balance equation, but this drift-diffusion-based treatment has very limited applicability in short-channel FETs. In the extended CPM, carrier energy is evaluated taking into account the band bending profile in the transport direction and the energy dissipation parameters such as the carrier mean free path and energy loss due to scattering.

The extended CPM was validated against HCD data acquired from foundry-quality nFETs (with secondary carriers being holes). Within model validation, the contribution of secondary holes was shown to be weak at the worst-case conditions for HCD ($V_{\text{gs}} = V_{\text{ds}}$) but became very significant at lower V_{gs}. This trend stems from two reasons: (1) the II rate, which generates the secondary carriers, is at its maximum when $V_{\text{gs}} \sim 0.5\ V_{\text{ds}}$ (which is shifted from the WCC), and (2) under the WCC, the secondary hole contribution is screened by the damage generated by primary electrons. It has also been shown that the hole-induced portion of HCD becomes stronger at higher V_{ds} values. This is because at a higher V_{ds} primary carriers have higher energies, thereby resulting in a higher II rate and a higher concentration of generated electron-hole pairs; in addition, secondary holes themselves can reach higher energies under an increased V_{ds}. Note that the interface trap density peak caused by secondary holes is located near the source, which is consistent with the results obtained with the full TCAD version of our HCD model. Finally, the extended CPM model was shown to accurately capture $\Delta I_{\text{d,lin}}(t)$ traces over a broad range of stress conditions.

Author Contributions: Conceptualization, S.T., E.B., A.G., Z.Y., A.M., A.D.K., M.V. (Mikhail Vexler), M.V. (Michiel Vandemaele), R.W., A.S., A.C. and B.K.; methodology, S.T., E.B., A.G., Z.Y., A.M., A.D.K., M.V. (Mikhail Vexler), M.V. (Michiel Vandemaele), R.W., A.S., A.C. and B.K.; software, S.T., E.B., A.G., Z.Y., A.M. and A.D.K.; validation, S.T.; formal analysis, S.T., E.B., A.G., Z.Y., A.M., A.D.K., M.V. (Mikhail Vexler), M.V. (Michiel Vandemaele), R.W., A.S., A.C. and B.K.; investigation, S.T., E.B., A.G., Z.Y., A.M., A.D.K., M.V. (Mikhail Vexler), M.V. (Michiel Vandemaele), R.W., A.S., A.C. and B.K.; resources, R.W., A.S. and A.C.; data curation, S.T., E.B, A.G. and B.K.; writing—original draft preparation, S.T.; writing—review and editing, S.T., E.B., A.G., Z.Y., A.M., A.D.K., M.V. (Mikhail Vexler), M.V. (Michiel Vandemaele), R.W., A.S., A.C. and B.K.; visualization, S.T. and A.M.;

supervision, R.W. and B.K.; project administration, R.W., A.S., A.C. and B.K.; funding acquisition, R.W., A.S. and A.C. All authors have read and agreed to the published version of the manuscript.

Funding: M.V. (Mikhail Vexler) acknowledges the support of the Ministry of Science and Higher Education of the Russian Federation (project no. 075-15-2020-790).

Data Availability Statement: The data presented in this study are available on request from the corresponding author.

Conflicts of Interest: The authors declare no conflict of interest.

Abbreviations

The following abbreviations are used in this manuscript:

BTE	Boltzmann Transport Equation
BTI	Bias Temperature Instability
CPM	Compact Physics Model
HCD	Hot-Carrier Degradation
II	Impact Ionization
MC	Multiple-Carrier (mechanism of Si-H bond dissociation)
MOSFET	Metal-Oxide-Semiconductor Field Effect Transistor
SC	Single-Carrier (mechanism of Si-H bond dissociation)
WCC	Worst-Case Conditions (of hot-carrier degradation)

References

1. Ramey, S.; Ashutosh, A.; Auth, C.; Clifford, J.; Hattendorf, M.; Hicks, J.; James, R.; Rahman, A.; Sharma, V.; Amour, A.S.; et al. Intrinsic transistor reliability improvements from 22 nm tri-gate technology. In Proceedings of the 2013 IEEE International Reliability Physics Symposium (IRPS), Monterey, CA, USA, 14–18 April 2013; pp. 4C.5.1–4C.5.5. [CrossRef]
2. Rahman, A.; Dacuna, J.; Nayak, P.; Leatherman, G.; Ramey, S. Reliability Studies of a 10 nm High-performance and Low-power CMOS Technology Featuring 3rd Generation FinFET and 5th Generation HK/MG. In Proceedings of the 2018 IEEE International Reliability Physics Symposium (IRPS), Burlingame, CA, USA, 11–15 March 2018; pp. 6F.4-1–6F.4-6. [CrossRef]
3. Paliwoda, P.; Chbili, Z.; Kerber, A.; Nigam, T.; Nagahiro, K.; Cimino, S.; Toledano-Luque, M.; Pantisano, L.; Min, B.W.; Misra, D. Self-Heating Effects on Hot Carrier Degradation and Its Impact on Logic Circuit Reliability. *IEEE Trans. Device Mater. Reliab.* **2019**, *19*, 249–254. [CrossRef]
4. McMahon, W.; Matsuda, K.; Lee, J.; Hess, K.; Lyding, J. The Effects of a Multiple Carrier Model of Interface States Generation of Lifetime Extraction for MOSFETs. In Proceedings of the International Conference on Modeling and Simulation of Microsystem, San Juan, PR, USA, 23–25 April 2002; Volume 1, pp. 576–579.
5. Haggag, A.; McMahon, W.; Hess, K.; Cheng, K.; Lee, J.; Lyding, J. High-performance Chip Reliability from Short-time-tests. Statistical Models for Optical Interconnect and HCI/TDDB/NBTI Deep-Submicron Transistor Failures. In Proceedings of the International Reliability Physics Symposium (IRPS), Orlando, FL, USA, 30 April–3 May 2001; pp. 271–279.
6. Reggiani, S.; Barone, G.; Poli, S.; Gnani, E.; Gnudi, A.; Baccarani, G.; Chuang, M.Y.; Tian, W.; Wise, R. TCAD Simulation of Hot-Carrier and Thermal Degradation in STI-LDMOS Transistors. *IEEE Trans. Electron Devices* **2013**, *60*, 691–698. [CrossRef]
7. Reggiani, S.; Barone, G.; Gnani, E.; Gnudi, A.; Baccarani, G.; Poli, S.; Wise, R.; Chuang, M.Y.; Tian, W.; Pendharkar, S.; et al. Characterization and Modeling of Electrical Stress Degradation in STI-based Integrated Power Devices. *Solid-State Electron.* **2014**, *102*, 25–41. [CrossRef]
8. Randriamihaja, Y.; Federspiel, X.; Huard, V.; Bravaix, A.; Palestri, P. New Hot Carrier Degradation Modeling Reconsidering the Role of EES in Ultra Short n-channel MOSFETs. In Proceedings of the International Reliability Physics Symposium (IRPS), Monterey, CA, USA, 14–18 April 2013; pp. 1–5.
9. Tyaginov, S.; Bina, M.; Franco, J.; Osintsev, D.; Triebl, O.; Kaczer, B.; Grasser, T. Physical Modeling of Hot-Carrier Degradation for Short- and Long-Channel MOSFETs. In Proceedings of the International Reliability Physics Symposium (IRPS), Waikoloa, HI, USA, 1–5 June 2014; pp. XT.16.1–XT.16.8.
10. Zaka, A.; Rafhay, Q.; Iellina, M.; Palestri, P.; Clerc, R.; Rideau, D.; Garetto, D.; Singer, J.; Pananakakis, G.; Tavernier, C.; et al. On the accuracy of current TCAD hot carrier injection models in nanoscale devices. *Solid State Electron.* **2010**, *54*, 1669–1674. [CrossRef]
11. Zaka, A.; Palestri, P.; Rafhay, Q.; Clerc, R.; Iellina, M.; Rideau, D.; Tavernier, C.; Pananakakis, G.; Jaouen, H.; Selmi, L. An Efficient Nonlocal Hot Electron Model Accounting for Electron-Electron Scattering. *IEEE Trans. Electron Devices* **2012**, *59*, 983–993. [CrossRef]
12. Hu, C.; Tam, S.; Hsu, F.; Ko, P.K.; Chan, T.Y.; Terrill, K. Hot-electron-induced MOSFET Degradation Model, Monitor and Improvement. *IEEE Trans. Electron Devices* **1985**, *48*, 375–385.

13. Hu, C.; Tam, S.; Hsu, F.C.; Ko, P.K.; Chan, T.Y.; Terrill, K. Hot-electron-induced MOSFET degradation—Model, monitor, and improvement. *IEEE J. Solid State Circuits* **1985**, *20*, 295–305.
14. Moens, P.; van den Bosch, G.; Groeseneken, G. Competing Hot Carrier Degradation Mechanisms in Lateral n-type DMOS Transistors. In Proceedings of the International Reliability Physics Symposium (IRPS), Dallas, TX, USA, 30 March–4 April 2003; pp. 214–221.
15. Moens, P.; Bauwens, F.; Nelson, M.; Tack, M. Electron trapping and interface trap generation in drain extended pMOS transistors. In Proceedings of the International Reliability Physics Symposium (IRPS), San Jose, CA, USA, 17–21 April 2005; pp. 93–96.
16. Moens, P.; van den Bosch, G. Characterization of total self-operating area of lateral DMOS transistors. *IEEE Trans Electron Devices* **2006**, *6*, 349–357.
17. Penzin, O.; Haggag, A.; McMahon, W.; Lyumkis, E.; Hess, K. MOSFET Degradation Kinetics and Its Simulation. *IEEE Trans. Electron Devices* **2003**, *50*, 1445–1450. [CrossRef]
18. Okhonin, S.; Hessler, T.; Dutoit, M. Comparison of gate-induced drain leakage and charge pumping measurements for determining lateral interface trap profiles in electrically stressed MOSFET's. *IEEE Trans. Electron Devices* **1996**, *43*, 605–612. [CrossRef]
19. Rauch, S.; La Rosa, G.; Guarin, F. Role of E-E Scattering in the Enhancement of Channel Hot Carrier Degradation of Deep-Submicron NMOSFETs at high V_{gs} Conditions. *IEEE Trans. Device Material. Reliab.* **2001**, *1*, 113–119. [CrossRef]
20. Rauch, S.; Rosa, G.L. The Energy Driven Paradigm of NMOSFET Hot Carrier Effects. In Proceedings of the International Reliability Physics Symposium (IRPS), San Jose, CA, USA, 17–21 April 2005.
21. Rauch, S.E.; Rosa, G.L. The energy-driven paradigm of NMOSFET hot-carrier effects. *IEEE Trans. Device Mater. Reliab.* **2005**, *5*, 701–705. [CrossRef]
22. Tyaginov, S.; Grill, A.; Vandemaele, M.; Grasser, T.; Hellings, G.; Makarov, A.; Jech, M.; Linten, D.; Kaczer, B. A Compact Physics Analytical Model for Hot-Carrier Degradation. In Proceedings of the 2020 IEEE International Reliability Physics Symposium (IRPS), Dallas, TX, USA, 28 April–30 May 2020; pp. 1–7. [CrossRef]
23. Rzepa, G.; Franco, J.; O'Sullivan, B.; Subirats, A.; Simicic, M.; Hellings, G.; Weckx, P.; Jech, M.; Knobloch, T.; Waltl, M.; et al. Comphy—A compact-physics framework for unified modeling of BTI. *Microelectron. Reliab.* **2018**, *85*, 49–65. [CrossRef]
24. Li, E.; Rosenbaum, E.; Tao, J.; Yeap, G.F.; Lin, M.; Fang, P. Hot-carrier Effects in nMOSFETs in 0.1 μm CMOS Technology. In Proceedings of the International Reliability Physics Symposium (IRPS), San Diego, CA, USA, 22–25 March 1999; pp. 253–258.
25. Lin, C.; Biesemans, S.; Han, L.; Houlihan, K.; Schiml, T.; Schruefer, K.; Wann, C.; Markhopf, R. Hot Carrier Reliability for 0.13 μm CMOS Technology with Dual Gate Oxide Thickness. In Proceedings of the International Electron Devices Meeting (IEDM), San Francisco, CA, USA, 10–13 December 2000; pp. 135–138.
26. Woltjer, R.; Hamada, A.; Takeda, E. PMOSFET Hot Carrier Damage: Oxide Charge and Interface States. *Semicond Sci. Technol.* **1992**, *7*, B581–B584. [CrossRef]
27. Tyaginov, S.; Starkov, I.; Enichlmair, H.; Park, J.; Jungemann, C.; Grasser, T. Physics-Based Hot-Carrier Degradation Models (invited). *ECS Trans.* **2011**, *35*, 321–352. [CrossRef]
28. Lee, K.; Kaczer, B.; Kruv, A.; Gonzalez, M.; Degraeve, R.; Tyaginov, S.; Grill, A.; De Wolf, I. Hot-Electron-Induced Punch-Through (HEIP) Effect in p-MOSFET Enhanced by Mechanical Stress. *IEEE Electron Device Lett.* **2021**, *42*, 1424–1427. [CrossRef]
29. Vandemaele, M.; Kaczer, B.; Tyaginov, S.; Stanojević, Z.; Makarov, A.; Chasin, A.; Bury, E.; Mertens, H.; Linten, D.; Groeseneken, G. Full (V_g, V_d) Bias Space Modeling of Hot-Carrier Degradation in Nanowire FETs. In Proceedings of the 2019 IEEE International Reliability Physics Symposium (IRPS), Monterey, CA, USA, 31 March–4 April 2019; pp. 1–7. [CrossRef]
30. Jech, M.; Tyaginov, S.; Kaczer, B.; Franco, J.; Jabs, D.; Jungemann, C.; Waltl, M.; Grasser, T. First–Principles Parameter–Free Modeling of n– and p–FET Hot–Carrier Degradation. In Proceedings of the 2019 IEEE International Electron Devices Meeting (IEDM), San Francisco, CA, USA, 7–11 December 2019; pp. 24.1.1–24.1.4. [CrossRef]
31. Vuillaume, D.; Marchetaux, J.C.; Lippens, P.E.; Bravaix, A.; Boudou, A. A coupled study by floating-gate and charge-pumping techniques of hot carrier-induced defects in submicrometer LDD n-MOSFET's. *IEEE Trans. Electron Devices* **1993**, *40*, 773–781. [CrossRef]
32. Chen, J.; Chen, S.Y.; Wu, K.M.; Liu, C. Investigation of hot-carrier-induced degradation mechanisms in p-type high-voltage drain extended metal-oxide-semiconductor transistors. *Jpn. J. Appl. Phys.* **2009**, *48*, 04C039. [CrossRef]
33. Starkov, I.; Enichlmair, H.; Tyaginov, S.; Grasser, T. Analysis of the Threshold Voltage Turn-Around Effect in High-Voltage n-MOSFETs due to Hot-Carrier Stress. In Proceedings of the International Reliability Physics Symposium (IRPS), Anaheim, CA, USA, 15–19 April 2012; p. 6.
34. Starkov, I.; Starkov, A. Investigation of the threshold voltage turn-around effect in long-channel n-MOSFETs due to hot-carrier stress. *Microelectron. Reliab.* **2014**, *54*, 33–36. [CrossRef]
35. Jung, S.G.; Lee, S.H.; Kim, C.K.; Yoo, M.S.; Yu, H.Y. Analysis of Drain Linear Current Turn-Around Effect in Off-State Stress Mode in pMOSFET. *IEEE Electron Device Lett.* **2020**, *41*, 804–807. [CrossRef]
36. Cho, M.; Spessot, A.; Kaczer, B.; Aoulaiche, M.; Ritzenthaler, R.; Schram, T.; Fazan, P.; Horiguchi, N.; Linten, D. Off-state stress degradation mechanism on advanced p-MOSFETs. In Proceedings of the 2015 International Conference on IC Design Technology (ICICDT), Leuven, Belgium, 1–3 June 2015; pp. 1–4. [CrossRef]

37. Spessot, A.; Aoulaiche, M.; Cho, M.; Franco, J.; Schram, T.; Ritzenthaler, R.; Kaczer, B. Impact of Off State Stress on advanced high-K metal gate NMOSFETs. In Proceedings of the 2014 44th European Solid State Device Research Conference (ESSDERC), Venice, Italy, 22–26 September 2014; pp. 365–368. [CrossRef]
38. Jech, M.; Rott, G.A.; Reisinger, H.; Tyaginov, S.E.; Rzepa, G.; Grill, A.; Jabs, D.; Jungemann, C.; Waltl, M.; Grasser, T. Mixed Hot-Carrier/Bias Temperature Instability Degradation Regimes in Full $\{V_G, V_D\}$ Bias Space: Implications and Peculiarities. *IEEE Trans. Electron Devices* **2020**, *67*, 3315–3322. [CrossRef]
39. Gritsch, M. Numerical Modeling of Silicon-on-Insulator MOSFETs. Ph.D. Thesis, Technische Universität, Wien, Austria, 2002.
40. Schenk, A. *Advanced Physical Models for Silicon Device Simulations*; Springer: Vienna, Austria, 1998.
41. Lundstrom, M. Drift-diffusion and computational electronics—Still going strong after 40 years! In Proceedings of the 2015 International Conference on Simulation of Semiconductor Processes and Devices (SISPAD), Washington, DC, USA, 9–11 September 2015; pp. 1–3. [CrossRef]
42. Tyaginov, S.; Bury, E.; Grill, A.; Yu, Z.; Makarov, A.; De Keersgieter, A.; Vexler, M.; Vandemaele, M.; Wang, R.; Spessot, A.; et al. On The Contribution of Secondary Holes in Hot-Carrier Degradation—A Compact Physics Modeling Perspective. In Proceedings of the 7th IEEE Electron Devices Technology & Manufacturing Conference (EDTM), Seoul, Republic of Korea, 7–10 March 2023; pp. 1–3. [CrossRef]
43. Auth, C.; Allen, C.; Blattner, A.; Bergstrom, D.; Brazier, M.; Bost, M.; Buehler, M.; Chikarmane, V.; Ghani, T.; Glassman, T.; et al. A 22 nm high performance and low-power CMOS technology featuring fully-depleted tri-gate transistors, self-aligned contacts and high density MIM capacitors. In Proceedings of the 2012 Symposium on VLSI Technology (VLSIT), Honolulu, HI, USA, 12–14 June 2012; pp. 131–132. [CrossRef]
44. Ferain, I.; Colinge, C.; Colinge, J.-P. Multigate transistors as the future of classical metal-oxide-semiconductor field-effect transistors. *Nature* **2011**, *479*, 310–316. [CrossRef] [PubMed]
45. Colinge, J.-P.; Lee, C.-W.; Afzalian, A.; Akhavan, N.D.; Yan, R.; Ferain, I.; Razavi, P.; O'Neill, B.; Blake, A.; White, M.; et al. Nanowire transistors without junctions. *Nat. Nano* **2010**, *5*, 225–229. . [CrossRef] [PubMed]
46. Kuhn, K.J. Considerations for Ultimate CMOS Scaling. *IEEE Trans. Electron Devices* **2012**, *59*, 1813–1828. [CrossRef]
47. Barraud, S.; Lapras, V.; Previtali, B.; Samson, M.P.; Lacord, J.; Martinie, S.; Jaud, M.A.; Athanasiou, S.; Triozon, F.; Rozeau, O.; et al. Performance and design considerations for gate-all-around stacked-NanoWires FETs. In Proceedings of the 2017 IEEE International Electron Devices Meeting (IEDM), San Francisco, CA, USA, 2–6 December 2017; pp. 29.2.1–29.2.4. [CrossRef]
48. Nagy, D.; Espiñeira, G.; Indalecio, G.; García-Loureiro, A.J.; Kalna, K.; Seoane, N. Benchmarking of FinFET, Nanosheet, and Nanowire FET Architectures for Future Technology Nodes. *IEEE Access* **2020**, *8*, 53196–53202. [CrossRef]
49. Weckx, P.; Ryckaert, J.; Putcha, V.; De Keersgieter, A.; Boemmels, J.; Schuddinck, P.; Jang, D.; Yakimets, D.; Bardon, M.G.; Ragnarsson, L.Å.; et al. Stacked nanosheet fork architecture for SRAM design and device co-optimization toward 3 nm. In Proceedings of the 2017 IEEE International Electron Devices Meeting (IEDM), San Francisco, CA, USA, 2–6 December 2017; pp. 20.5.1–20.5.4. [CrossRef]
50. Zhang, J.; Frougier, J.; Greene, A.; Miao, X.; Yu, L.; Vega, R.; Montanini, P.; Durfee, C.; Gaul, A.; Pancharatnam, S.; et al. Full Bottom Dielectric Isolation to Enable Stacked Nanosheet Transistor for Low Power and High Performance Applications. In Proceedings of the 2019 IEEE International Electron Devices Meeting (IEDM), San Francisco, CA, USA, 7–11 December 2019; pp. 11.6.1–11.6.4. [CrossRef]
51. Ritzenthaler, R.; Mertens, H.; Eneman, G.; Simoen, E.; Bury, E.; Eyben, P.; Bufler, F.M.; Oniki, Y.; Briggs, B.; Chan, B.; et al. Comparison of Electrical Performance of Co-Integrated Forksheets and Nanosheets Transistors for the 2 nm Technological Node and Beyond. In Proceedings of the 2021 IEEE International Electron Devices Meeting (IEDM), San Francisco, CA, USA, 13–15 December 2021; pp. 26.2.1–26.2.4. [CrossRef]
52. Ryckaert, J.; Schuddinck, P.; Weckx, P.; Bouche, G.; Vincent, B.; Smith, J.; Sherazi, Y.; Mallik, A.; Mertens, H.; Demuynck, S.; et al. The Complementary FET (CFET) for CMOS scaling beyond N3. In Proceedings of the 2018 IEEE Symposium on VLSI Technology, Honolulu, HI, USA, 18–22 June 2018; pp. 141–142. [CrossRef]
53. Liebmann, L.; Smith, J.; Chanemougame, D.; Gutwin, P. CFET Design Options, Challenges, and Opportunities for 3D Integration. In Proceedings of the 2021 IEEE International Electron Devices Meeting (IEDM), San Francisco, CA, USA, 13–15 December 2021; pp. 3.1.1–3.1.4. [CrossRef]
54. Bury, E.; Kaczer, B.; Chuang, K.; Franco, J.; Weckx, P.; Chasin, A.; Simicic, M.; Linten, D.; Groeseneken, G. Statistical assessment of the full VG/VD degradation space using dedicated device arrays. In Proceedings of the 2017 IEEE International Reliability Physics Symposium (IRPS), Monterey, CA, USA, 2–6 April 2017; pp. 2D–5.1–2D–5.6. [CrossRef]
55. Grill, A.; Bury, E.; Michl, J.; Tyaginov, S.; Linten, D.; Grasser, T.; Parvais, B.; Kaczer, B.; Waltl, M.; Radu, I. Reliability and Variability of Advanced CMOS Devices at Cryogenic Temperatures. In Proceedings of the International Reliability Physics Symposium (IRPS), Dallas, TX, USA, 28 April–30 May 2020; pp. 5C.3.1–5C.3.1.6.
56. Krick, J.; Lenahan, P.; Dunn, G. Direct observation of interfacial point defects generated by channel hot hole injection in nchannel metal oxide silicon field effect transistors. *Appl. Phys. Lett.* **1991**, *59*, 3437–3439. [CrossRef]
57. Billman, C.; Lenahan, P.; Weber, W. Identification of the Microscopic Structure of New Hot Carrier Damage in Short Channel MOSFETs. *Microelectron. Eng.* **1997**, *36*, 271–272. [CrossRef]
58. Lenahan, P. Atomic Scale Defects Involved in MOS Reliability Problems. *Microelectron. Eng.* **2003**, *69*, 173–181. [CrossRef]

59. Helms, C.R.; Poindexter, E.H. The Silicon-Silicon Dioxide System: Its Microstructure and Imperfections. *Rep. Prog. Phys.* **1994**, *57*, 791. [CrossRef]
60. Stesmans, A. Passivation of P_{b0} and P_{b1} Interface Defects in Thermal (100) Si/SiO_2 with Molecular Hydrogen. *Appl. Phys. Lett.* **1996**, *68*, 2076–2078. [CrossRef]
61. Pobegen, G.; Tyaginov, S.; Nelhiebel, M.; Grasser, T. Observation of Normally Distributed Activation Energies for the Recovery from Hot Carrier Damage. *IEEE Electron Device Lett.* **2013**, *34*, 939–941. [CrossRef]
62. de Jong, M.J.; Salm, C.; Schmitz, J. Towards Understanding Recovery of Hot-Carrier Induced Degradation. *Microelectron. Reliab.* **2018**, *88–90*, 147–151. [CrossRef]
63. de Jong, M.J.; Salm, C.; Schmitz, J. Effect of Ambient on the Recovery of Hot-Carrier Degraded Devices. In Proceedings of the 2020 IEEE International Reliability Physics Symposium (IRPS), Dallas, TX, USA, 28 April–30 May 2020; pp. 1–6.
64. Vandemaele, M.; Chuang, K.H.; Bury, E.; Tyaginov, S.; Groeseneken, G.; Kaczer, B. The Influence of Gate Bias on the Anneal of Hot-Carrier Degradation. In Proceedings of the International Reliability Physics Symposium (IRPS), Dallas, TX, USA, 28 April–30 May 2020; pp. 5A.3.1–5A.3.7. [CrossRef]
65. Reisinger, H.; Blank, O.; Heinrigs, W.; Muhlhoff, A.; Gustin, W.; Schlunder, C. Analysis of NBTI Degradation- and Recovery-Behavior Based on Ultra Fast VT-Measurements. In Proceedings of the 2006 IEEE International Reliability Physics Symposium, San Jose, CA, USA, 26–30 March 2006; pp. 448–453. [CrossRef]
66. Grasser, T.; Kaczer, B.; Gös, W.; Reisinger, H.; Aichinger, T.; Hehenberger, P.; Wagner, P.J.; Franco, J.; Toledano-Luque, M.; Nelhiebel, M. The Paradigm Shift in Understanding the Bias Temperature Instability: From Reaction-Diffusion to Switching Oxide Traps. *IEEE Trans. Electron Devices* **2011**, *58*, 3652–3666. [CrossRef]
67. Grasser, T.; Kaczer, B.; Goes, W.; Reisinger, H.; Aichinger, T.; Hehenberger, P.; Wagner, P.J.; Schanowsky, F.; Franco, J.; Roussel, P.; et al. Recent Advances in Understanding the Bias Temperature Instability. In Proceedings of the International Electron Devices Meeting (IEDM), San Francisco, CA, USA, 6–8 December 2010; pp. 82–85.
68. Bravaix, A.; Guerin, C.; Huard, V.; Roy, D.; Roux, J.; Vincent, E. Hot-carrier Acceleration Factors for Low Power Management in DC-AC Stressed 40 nm NMOS Node at High Temperature. In Proceedings of the International Reliability Physics Symposium (IRPS), Montreal, QC, Canada, 26–30 April 2009; pp. 531–546. [CrossRef]
69. Hess, K.; Register, L.; Tuttle, B.; Lyding, J.; Kizilyalli, I. Impact of Nanostructure Research on Conventional Solid-State Electronics: The Giant Isotope Effect in Hydrogen Desorption and CMOS Lifetime. *Physica E* **1998**, *3*, 1–7. [CrossRef]
70. Sharma, P.; Tyaginov, S.; Jech, M.; Wimmer, Y.; Rudolf, F.; Enichlmair, H.; Park, J.M.; Ceric, H.; Grasser, T. The role of cold carriers and the multiple-carrier process of Si–H bond dissociation for hot-carrier degradation in n- and p-channel {LDMOS} devices. *Solid State Electron.* **2016**, *115 Pt B*, 185–191. [CrossRef]
71. Tallarico, A.N.; Reggiani, S.; Depetro, R.; Torti, A.M.; Croce, G.; Sangiorgi, E.; Fiegna, C. Hot-Carrier Degradation in Power LDMOS: Selective LOCOS- Versus STI-Based Architecture. *IEEE J. Electron Devices Soc.* **2018**, *6*, 219–226. [CrossRef]
72. Kerber, A.; Nigam, T.; Paliwoda, P.; Guarin, F. Reliability Characterization of Ring Oscillator Circuits for Advanced CMOS Technologies. *IEEE Trans. Device Mater. Reliab.* **2020**, *20*, 230–241. [CrossRef]
73. Bravaix, A.; Huard, V.; Goguenheim, D.; Vincent, E. Hot-Carrier to Cold-Carrier Device Lifetime Modeling with Temperature for Low power 40 nm Si-Bulk NMOS and PMOS FETs. In Proceedings of the International Electron Devices Meeting (IEDM), Washington, DC, USA, 5–7 December 2011; pp. 622–625.
74. Bina, M.; Tyaginov, S.; Franco, J.; Rupp, K.; Wimmer, Y.; Osinstev, D.; Kaczer, B.; Grasser, T. Predictive Hot-Carrier Modeling of n-channel MOSFETs. *IEEE Trans. Electron Devices* **2014**, *61*, 3103–3110. [CrossRef]
75. Varghese, D.; Alam, M.A.; Weir, B. A generalized, IB-independent, physical HCI lifetime projection methodology based on universality of hot-carrier degradation. In Proceedings of the 2010 IEEE International Reliability Physics Symposium, Anaheim, CA, USA, 2–6 May 2010; pp. 1091–1094. [CrossRef]
76. Yamagata, M.; Satoh, T.; Tango, H. Hot-Carrier-Induced Degradation under Current Saturation Bias in p-Channel Low-Temperature Polycrystalline Silicon Thin-Film Transistors. *JPN J. Appl. Phys.* **2007**, *46*, 5044. [CrossRef]
77. Makarov, A.; Tyaginov, S.E.; Kaczer, B.; Jech, M.; Chasin, A.; Grill, A.; Hellings, G.; Vexler, M.I.; Linten, D.; Grasser, T. Hot-Carrier Degradation in FinFETs: Modeling, Peculiarities, and Impact of Device Topology. In Proceedings of the 2017 IEEE International Electron Devices Meeting (IEDM), San Francisco, CA, USA, 2–6 December 2017; pp. 13.1.1–13.1.4. [CrossRef]
78. Tyaginov, S.; Jech, M.; Franco, J.; Sharma, P.; Kaczer, B.; Grasser, T. Understanding and Modeling the Temperature Behavior of Hot-Carrier Degradation in SiON nMOSFETs. *IEEE Electron Device Lett.* **2016**, *37*, 84–87. [CrossRef]
79. McMahon, W.; Haggag, A.; Hess, K. Reliability Scaling Issues for Nanoscale Devices. *IEEE Trans. Nanotech.* **2003**, *2*, 33–38. [CrossRef]
80. Haggag, A.; Lemanski, M.; Anderson, G.; Abramowitz, P.; Moosa, M. Realistic Projections of Product Fmax Shift and Statistics due to HCI and NBTI. In Proceedings of the 2007 IEEE 45th Annual International Reliability Physics Symposium, Phoenix, AZ, USA, 15–19 April 2007; pp. 93–96. [CrossRef]
81. ViennaSHE. 2014. Available online: http://viennashe.sourceforge.net/ (accessed on 5 October 2023).
82. Rupp, K.; Jungel, A.; Grasser, T. Matrix compression for spherical harmonics expansions of the Boltzmann transport equation for semiconductors. *J. Comput. Phys.* **2010**, *229*, 8750–8765. [CrossRef]

83. Rupp, K.; Grasser, T.; Jüngel, A. On the Feasibility of Spherical Harmonics Expansions of the Boltzmann Transport Equation for Three-Dimensional Device Geometries. In Proceedings of the 2011 International Electron Devices Meeting, Washington, DC, USA, 5–7 December 2011; pp. 34.1.1–34.1.4. [CrossRef]
84. Rupp, K.; Jungemann, C.; Hong, S.M.; Bina, M.; Grasser, T.; Jüngel, A. A Review of Recent Advances in the Spherical Harmonics Expansion Method for Semiconductor Device Simulation. *J. Comput. Electron.* **2016**, *15*, 939–958. [CrossRef]
85. Gnudi, A.; Ventura, D.; Baccarani, G. One-Dimensional Simulation of a Bipolar Transistor by means of Spherical Harmonics Expansion of the Boltzmann Transport Equation. In Proceedings of the Simulation of Semiconductor Devices and Processes (SISDEP), Zurich, Switzerland, 12–14 September 1991; Volume 4, pp. 205–213.
86. Hong, S.M.; Jungemann, C. A fully coupled scheme for a Boltzmann-Poisson equation solver based on a spherical harmonics expansion. *J. Comput. Electron.* **2009**, *8*, 225–241. [CrossRef]
87. Hong, S.M.; Pham, A.; Jungemann, C. *Deterministic Solvers for the Boltzmann Transport Equation*; Springer: Vienna, Austria, 2011.
88. Grasser, T.; Kosina, H.; Heitzinger, C.; Selberherr, S. Accurate Impact Ionization Model which Accounts for Hot and Cold Carrier Populations. *Appl. Phys. Lett.* **2002**, *80*, 613–615. [CrossRef]
89. Selmi, L.; Sangiorgi, E.; Bez, R.; Ricco, B. Measurement of the hot hole injection probability from Si into SiO_2 in p-MOSFETs. In Proceedings of the IEEE International Electron Devices Meeting, Washington, DC, USA, 5–8 December 1993; pp. 333–336. . [CrossRef]
90. Tyaginov, S.; Bina, M.; Franco, J.; Wimmer, Y.; Osintsev, D.; Kaczer, B.; Grasser, T. A Predictive Physical Model for Hot-Carrier Degradation in Ultra-Scaled MOSFETs. In Proceedings of the Simulation of Semiconductor Processes and Devices (SISPAD), Yokohama, Japan, 9–11 September 2014; pp. 89–92.
91. Vecchi, M.C.; Rudan, M. Modeling electron and hole transport with full-band structure effects by means of the Spherical-Harmonics Expansion of the BTE. *IEEE Trans. Electron Devices* **1998**, *45*, 230–238. [CrossRef]
92. Fischetti, M.V.; Yoder, P.D.; Khatami, M.M.; Gaddemane, G.; Van de Put, M.L. "Hot Electrons in Si Lose Energy Mostly to Optical Phonons": Truth or Myth? *Appl. Phys. Lett.* **2019**, *114*, 222104. [CrossRef]
93. Tandon, N.; Albrecht, J.D.; Ram-Mohan, L.R. Electron-Phonon Interaction and Scattering in Si and Ge: Implications for Phonon Engineering. *J. Appl. Phys.* **2015**, *118*, 045713. [CrossRef]
94. Qiu, B.; Tian, Z.; Vallabhaneni, A.; Liao, B.; Mendoza, J.M.; Restrepo, O.D.; Ruan, X.; Chen, G. First-Principles Simulation of Electron Mean-Free-Path Spectra and Thermoelectric Properties in Silicon. *Europhys. Lett.* **2015**, *109*, 57006. [CrossRef]
95. Jungemann, C.; Meinerzhagen, B. *Hierarchical Device Simulation*; Springer: Wien, Austria, 2003.
96. Jacoboni, C.; Lugli, P. *The Monte Carlo Method for Semiconductor Device Simulation*; Springer: Wien, Austria, 1989.
97. Sze, S.; Ng, K. *Physics of Semiconductor Devices*; Wiley-Interscience: Hoboken, NJ, USA, 2007.
98. Leblebici, Y.; Kang, S.M. Modeling of nMOS Transistors for Simulation of Hot-Carrier Induced Device and Circuit Degradation. *IEEE Trans. Comput. Aided Des.* **1992**, *11*, 235–246. [CrossRef]
99. Acovic, A.; Rosa, G.L.; Sun, Y. A Review of Hot Carrier Degradation Mechanism in MOSFETs. *Microel. Reliab.* **1996**, *36*, 845–869. [CrossRef]
100. Ancona, M.G.; Saks, N.S.; McCarthy, D. Lateral distribution of hot-carrier-induced interface traps in MOSFETs. *IEEE Trans. Electron Devices* **1988**, *35*, 2221–2228. [CrossRef]
101. Varghese, D.; Kufluoglu, H.; Reddy, V.; Shichijo, H.; Krishnan, S.; Alam, M.A. Universality of Off-State Degradation in Drain Extended NMOS Transistors. In Proceedings of the 2006 International Electron Devices Meeting, San Francisco, CA, USA, 11–13 December 2006; pp. 1–4. [CrossRef]

micromachines

MDPI

Article

Design of a Novel Compact Bandpass Filter Based on Low-Cost Through-Silicon-Via Technology

Hai Dong [1], Yingtao Ding [1], Han Wang [1], Xingling Pan [1], Mingrui Zhou [1] and Ziyue Zhang [1,2,*]

[1] School of Integrated Circuits and Electronics, Beijing Institute of Technology, Beijing 100081, China; 3420225025@bit.edu.cn (H.D.); ytd@bit.edu.cn (Y.D.); 3220215100@bit.edu.cn (H.W.); panxingling0705@163.com (X.P.); linghuaibu@163.com (M.Z.)
[2] BIT Chongqing Institute of Microelectronics and Microsystems, Chongqing 400030, China
* Correspondence: zyzhang@bit.edu.cn

Abstract: Three-dimensional (3D) integration based on through-silicon-via (TSV) technology provides a solution to the miniaturization of electronic systems. In this paper, novel integrated passive devices (IPDs) including capacitor, inductor, and bandpass filter are designed by employing TSV structures. For lower manufacturing costs, polyimide (PI) liners are used in the TSVs. The influences of structural parameters of TSVs on the electrical performance of the TSV-based capacitor and inductor are individually evaluated. Moreover, with the topologies of capacitor and inductor elements, a compact third-order Butterworth bandpass filter with a central frequency of 2.4 GHz is developed, and the footprint is only 0.814 mm × 0.444 mm. The simulated 3-dB bandwidth of the filter is 410 MHz, and the fraction bandwidth (FBW) is 17%. Besides, the in-band insertion loss is less than 2.63 dB, and the return loss in the passband is better than 11.4 dB, showing good RF performance. Furthermore, as the filter is fully formed by identical TSVs, it not only features a simple architecture and low cost, but also provides a promising idea for facilitating the system integration and layout camouflaging of radio frequency (RF) devices.

Keywords: compact bandpass filter; integrated passive device (IPD); miniaturization; three-dimensional (3D) integration; through-silicon-via (TSV)

Citation: Dong, H.; Ding, Y.; Wang, H.; Pan, X.; Zhou, M.; Zhang, Z. Design of a Novel Compact Bandpass Filter Based on Low-Cost Through-Silicon-Via Technology. *Micromachines* **2023**, *14*, 1251. https://doi.org/10.3390/mi14061251

Academic Editor: Niall Tait

Received: 18 May 2023
Revised: 10 June 2023
Accepted: 13 June 2023
Published: 14 June 2023

1. Introduction

As one of the key components in radio frequency (RF) front ends and microwave systems, bandpass filters (BPFs) are extensively investigated and have spurred great advances in terms of structures, processes, and performance over the past decades. Several technologies such as low temperature co-fired ceramic (LTCC) [1,2], microstrip [3–6], substrate integrated waveguide (SIW) [7–9], and micro-electromechanical systems (MEMS) [10,11], are utilized to develop BPFs with specific characteristics such as broad bandwidth, high selectivity, low insertion loss, and compact size. Recently, with the rapid development of wireless communication, Internet of Things (IoT), and Artificial Intelligence (AI), the demands for miniaturization and integration of RF components are especially urgent, therefore, it is necessary to design filters that are more compact and easier to be integrated [12–15].

To achieve a smaller footprint, multi-layer structure and discriminating coupling are involved in the design of compact LTCC BPFs [16–18]. For SIW filters, some modified topologies such as the half-mode SIW (HMSIW) [19], and the ridge half-mode SIW (RHMSIW) [20,21], are proposed to optimize the device size. Besides, based on the spoof surface plasmon polaritons (SSPPs) in the microwave frequencies, several novel compact wideband SSPP filters [22,23] and hybrid SIW-SSPP filters [24,25] are developed for plasmonic systems, microwave circuits, and wearable devices. MEMS technology offers another alternative to the design of compact integrated passive devices (IPDs) [26,27], and the RF MEMS devices are superior in aspects of tuning, switching, and integration with other chips [28,29]. However, it is difficult for most of the above technologies to further shrink the footprints of filters to satisfy the increasing demands for the miniaturization of future electronic systems.

Three-dimensional (3D) integration and interposer technologies have emerged as effective solutions to the heterogeneous integration of multiple devices and chips [30–32]. As the core of 3D integration and interposer technologies, through-silicon-vias (TSVs) can vertically transmit electrical signals among different device layers, enabling high-performance systems within more compact areas [33–35]. Moreover, TSVs can also constitute several functional devices, such as capacitors [36] and inductors [37,38]. For RF applications, glass substrate is also used due to its high electrical resistivity and low electrical loss, where through-glass-vias (TGVs) are developed similar to TSVs [39,40]. Some of the reported literature has used TSV/TGV based capacitors together with inductors fabricated by traditional integrated circuit (IC) processes (or vice versa) to form BPF structures [41–44], showing great potential to improve the compactness of filters. To further simplify the device structure and reduce the process complexity, a compact, fully TGV-based BPF is proposed in [45]. Nevertheless, the manufacture of TGVs is not as mature as that of TSVs, which is more flexible and reliable in fabrication processes such as via etching and metallization. On the other hand, TSV technology has been widely studied and has achieved various progresses, for example, some low dielectric constant polymers are developed as substitutes for conventional SiO_2 liners, which feature low parasitic capacitance, good insulation quality, and low fabrication cost and complexity [46–48]. In this paper, novel and compact IPD structures including capacitor, inductor, and BPF are designed based on TSV technology. High resistivity silicon (HRS) with a resistivity of 1×10^4 Ω·cm is used as the substrate has higher RF performance, as well as a better capability to be integrated with other Si based devices. Besides, polyimide (PI) is applied as the TSV liners, which can offer good electrical performance and be easily fabricated by a low-cost and simple technique named vacuum-assisted spin coating [49–51]. The electrical characteristics of the TSV based coaxial-like capacitors and solenoid inductors are discussed in detail. Notably, the utilized TSVs are identical and thus can be fabricated in a single batch. Employing the TSV based capacitor and inductor elements, a compact BPF structure is designed and evaluated, which possesses several attractive advantages, particularly with respect to a small footprint, simple structure, and low cost.

This paper is organized as follows: Sections 2 and 3 discuss the influences of design parameters on the performance of the TSV based coaxial-like capacitors and the TSV based solenoid inductors, respectively. Section 4 describes the modeling and characterization of the proposed BPF. Finally, the conclusions are summarized in Section 5.

2. TSV Based Coaxial-like Capacitor

Traditional on-chip capacitors are usually planar metal-insulator-metal (MIM) structures occupying large device areas. By sequentially depositing multiple layers via holes, 3D MIM capacitors named through-silicon-capacitors (TSCs) are fabricated [52–54]. Even though the TSCs have a smaller footprint than traditional planar capacitors, their fabrication processes are relatively complex. Actually, the TSV structure is a natural metal-insulator-semiconductor (MIS) capacitor, although the capacitance is always regarded as an unwanted parasitic parameter. Besides, if one TSV is placed at the center of several annularly arranged TSVs that are connected together, a coaxial-like capacitor structure can be formed, where the central conductor of the inner TSV and the central conductors of the surrounding TSVs act as the two terminals of the capacitor [55], respectively.

Based on the low-cost PI-TSV technology, a topological coaxial-like capacitor structure is firstly proposed, and the basic capacitor element comprised of five identical TSVs is shown in Figure 1. The HRS substrate is hidden for better observation. The TSVs have a diameter of 30 μm and a height of 300 μm, the thickness of PI liner is 0.5 μm, and the pitch between inner and outer TSVs is $25\sqrt{2}$ μm as the pitch between adjacent outer TSVs is set to be 50 μm. It can be seen that four surrounding TSVs are connected to a port by a backside

Cu redistribution layer (RDL), and the central TSV is linked to the other port by a frontside RDL. Actually, more surrounding TSVs contributes to increasing the capacitance of the coaxial-like capacitor [55], however, to build capacitors that have specific capacitances, the number of surrounding TSVs is set to be four, forming a square shape that can be easily duplicated.

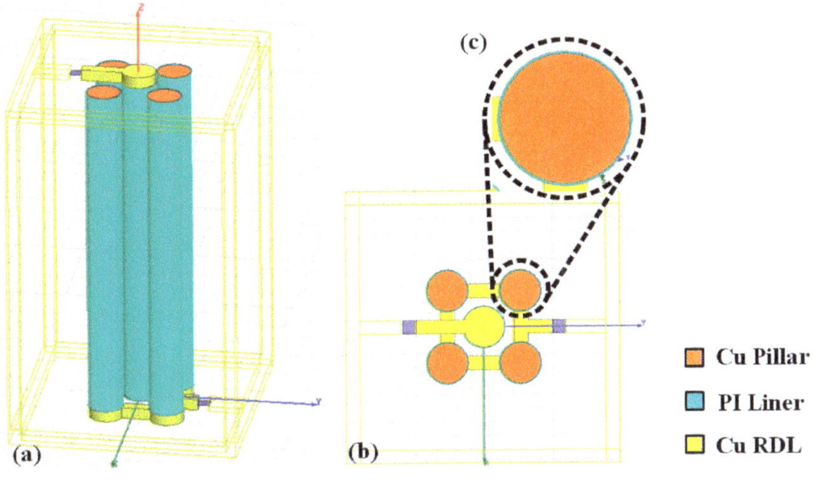

Figure 1. Basic capacitor element for the PI-TSV based coaxial-like capacitor structure: (**a**) side view, (**b**) top view, (**c**) is the enlarged image of a single TSV in (**b**).

Considering the sandwiched substrate between the inner and surrounding TSVs, the capacitance of such a coaxial-like capacitor is formed by the series connection of the parasitic capacitances of the inner TSV, the capacitance of the HRS substrate, and the parasitic capacitances of the surrounding TSVs. Therefore, the total capacitance of the capacitor element is influenced by the structural parameters of the TSVs and the pitch between central and outer TSVs.

Figure 2 plots the simulated capacitance–frequency characteristics of the capacitor element in Figure 1 under various design parameters, including TSV height, TSV diameter, thickness of PI liner, and pitch between inner and outer TSVs. As shown in Figure 2a,b, the increases in TSV height and diameter enlarge the capacitances over the simulated frequencies, due to the increased equivalent terminal areas in the capacitor. In addition, with the decreases in PI thickness and pitch between inner and outer TSVs, the parasitic capacitances of the PI-TSVs and the capacitance of the substrate are increased, respectively, leading to the increase in the total capacitance.

Moreover, arrays of coaxial-like capacitors with larger capacitances can be easily formed by the duplication and parallel connection of the basic square capacitor elements in both X- and Y-directions, which can be described as the matrix topology of coaxial-like capacitors.

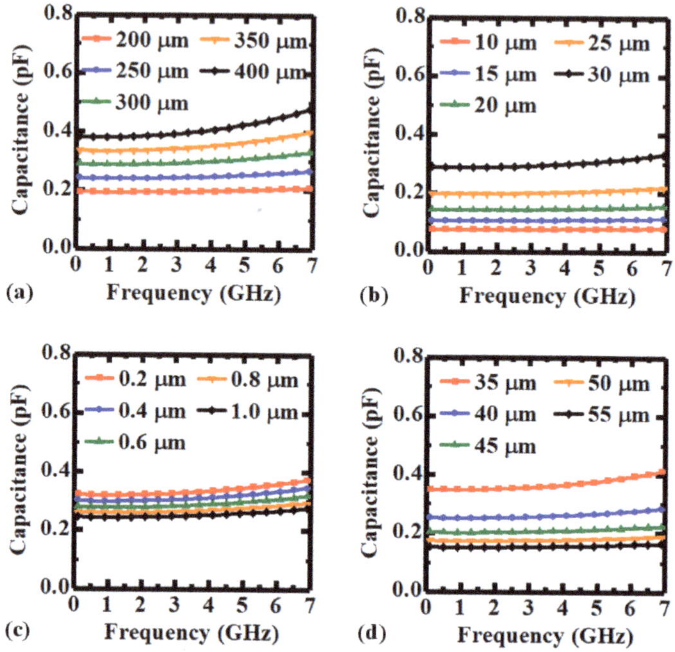

Figure 2. Capacitance–frequency characteristics of the PI-TSV based capacitor element with various (**a**) TSV heights, (**b**) TSV diameters, (**c**) thicknesses of PI liner, and (**d**) pitches between inner and outer TSVs.

3. TSV Based Solenoid Inductor

Three-dimensional (3D) solenoid inductors comprised of TSVs/TGVs and RDLs have been proposed for a smaller footprint and higher inductance density than traditional planar spiral structures [39,40]. To further reduce the manufacturing cost and complexity, a new 3D solenoid inductor structure based on low-cost PI-TSVs and corresponding RDLs is designed, and the basic inductor element with two turns is shown in Figure 3. The physical parameters of the PI-TSVs are the same as those of the coaxial-like capacitor in Section 2. It is shown that two X-directional adjacent TSVs and the RDLs form an equivalent coil, and the equivalent cross-sectional area of the inductor is decided by the pitch between the two TSVs and the TSV height. Actually, 3D solenoid inductors can store magnetic field energy more efficiently than planar structures, thus achieving larger quality factors within smaller footprints [39]. On the other hand, the pitch between two adjacent Y-directional TSVs acts as the loop pitch of the inductor, thus influencing the performance of the inductor together with the cross-sectional area and the number of turns. Therefore, the impact of various structural parameters including TSV height, TSV diameter, thickness of PI liner, and pitches between two adjacent TSVs in X- and Y-directions on the inductance and quality factor of the inductor element is analyzed, and the simulated results are shown in Figures 4 and 5.

Figure 4a shows that the inductance increases with TSV height as the cross-sectional area of the inductor increases. Besides, as the frequency-dependent alternating current (AC) resistance also increases with TSV height, the variation trend of the quality factor is determined by the weights of the increasing ratios of the reactance and the resistance at various frequencies. Therefore, the quality factor increases with TSV height at lower frequencies but decreases at higher frequencies, as shown in Figure 5a. As the increase in TSV diameter decreases the effective cross-sectional area, the inductance is declined accordingly, as shown in Figure 4b. However, as the resistance is decreased more significantly, the quality factor is increased instead, as shown in Figure 5b. Figures 4c and 5c illustrate that the thickness of PI liner has little impact on the inductance and quality factor of the inductor element, which is because both the effective cross-sectional area and the resistance are almost unchanged, and the reduction in energy loss with thicker liner is not significant. As shown in Figure 4d, the influence of the pitch between two X-directional TSVs on the inductance is similar to that of the TSV height as both of them directly decide the effective cross-sectional area of the inductor. Notably, as the pitch between the two TSVs has less impact on the resistance of the coil compared to the TSV height, the quality factor is increased with the inductance, as shown in Figure 5d. Finally, the pitch between two Y-directional adjacent TSVs is actually influencing the equivalent length of the inductor, therefore, the inductance is slightly decreased with larger pitch, as shown in Figure 4e. On the other hand, as the inductive coupling is also decreased due to proximity effect, the quality factor is nearly the same, as shown in Figure 5e.

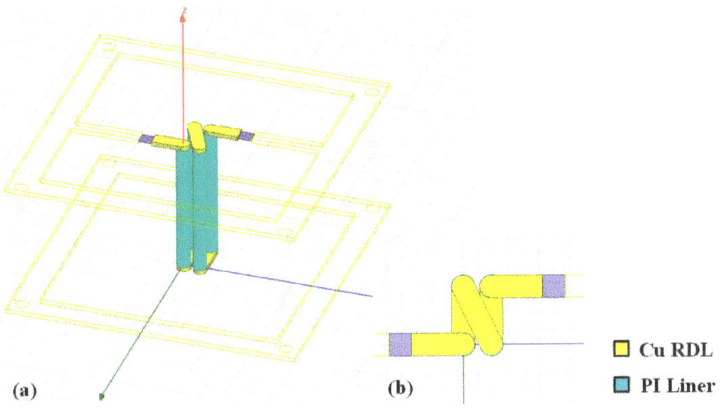

Figure 3. Basic inductor element for the PI-TSV based solenoid inductor structure: (**a**) side view, (**b**) top view.

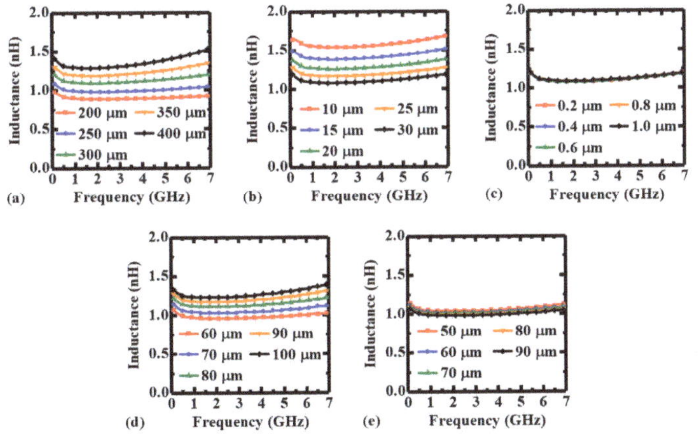

Figure 4. Inductance–frequency characteristics of the PI-TSV based inductor element with various (**a**) TSV heights, (**b**) TSV diameters, (**c**) thicknesses of PI liner, (**d**) pitch between two X-directional adjacent TSVs, and (**e**) pitch between two Y-directional adjacent TSVs.

Similar to the coaxial-like capacitor in Section 2, the basic solenoid inductor element can also constitute inductors with larger inductances in a simple topological method. As the proposed inductor element is composed of repeatable loops of identical TSVs and RDLs, an array of 2 × N TSVs can easily be formed to multiply the inductance, where N refers to the number of turns. Besides, the pitches between two adjacent TSVs in X- and Y-directions can also be adjusted to tune the inductance of the basic inductor element.

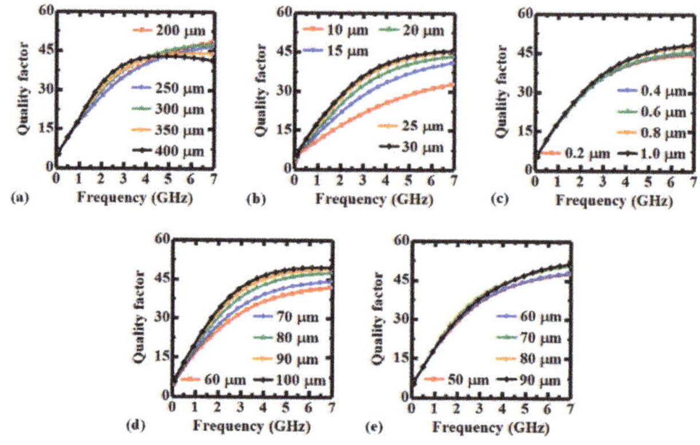

Figure 5. Quality factor-frequency characteristics of the PI-TSV based inductor element with various (**a**) TSV heights, (**b**) TSV diameters, (**c**) thicknesses of PI liner, (**d**) pitch between two X-directional adjacent TSVs, and (**e**) pitch between two Y-directional adjacent TSVs.

4. TSV Based Bandpass Filter

Based on the proposed coaxial-like capacitor and solenoid inductor structures, a novel BPF structure is designed, which is fully comprised of identical low-cost PI-TSVs. A typical prototype of third-order Butterworth BPF is adopted and the simplified circuit model of the designed filter with a central frequency (f_0) of 2.4 GHz is shown in Figure 6. Three capacitors and three inductors are used in the filter, and the parameter values of the components are derived and validated by the Keysight ADS circuit simulator. To obtain PI-TSV based capacitors and inductors of specific values, the coaxial-like capacitor element and the solenoid inductor element are topologically evolved.

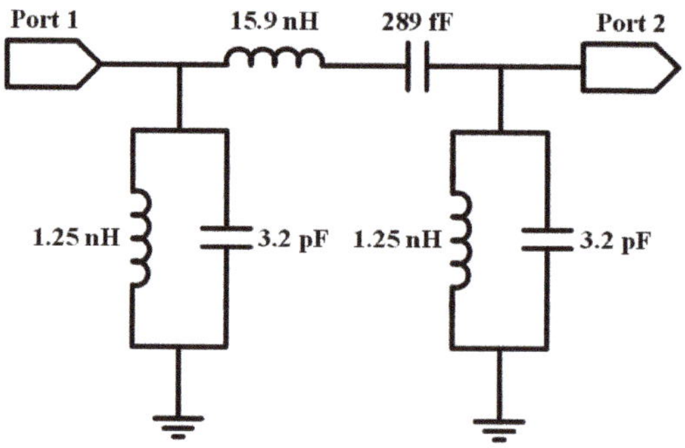

Figure 6. Circuit model of the proposed third-order Butterworth BPF.

Figure 7 shows the designed PI-TSV based BPF structure that is assembled by three arrays of coaxial-like capacitors and three 3D solenoid inductors with multi-turns. Notably, all TSVs in the filter are identical thus can be fabricated simultaneously, which greatly reduces the manufacturing cost and complexity. Moreover, as shown in Figure 7b, the footprint of the compact BPF is only 0.814 mm × 0.444 mm excluding the surrounding test framework, corresponding to an ultra-small electrical dimension of about 0.0065 λ_0 × 0.0035 λ_0.

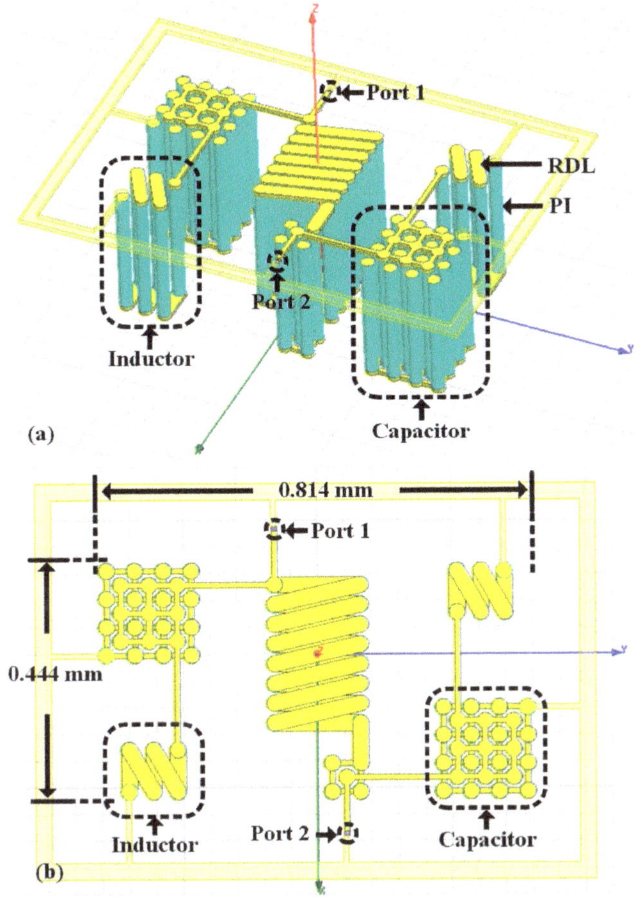

Figure 7. Structure of the proposed PI-TSV based BPF: (**a**) side view, (**b**) top view. A coaxial-like capacitor component and a solenoid inductor component are indicated. The device footprint is 0.814 mm × 0.444 mm.

The transmission characteristics of the proposed BPF are analyzed using full-wave simulations in Ansys HFSS, and the return loss (S_{11}) and insertion loss (S_{21}) curves are plotted in Figure 8. Besides, the S_{11} and S_{21} curves obtained from the theoretical circuit model are also shown in Figure 8 for comparison. According to the simulated results, the f_0 of the BPF is 2.4 GHz, the 3-dB bandwidth is 410 MHz, and the fraction bandwidth (FBW) is 17%. In addition, the in-band insertion loss is less than 2.63 dB, and the return loss in the passband is better than 11.4 dB, showing good RF performance. Note that there are some deviations between the S-parameter curves of the circuit model and the physical model of the proposed PI-TSV based filter, which can be attributed to the parasitic parameters introduced by the TSV and RDL structures.

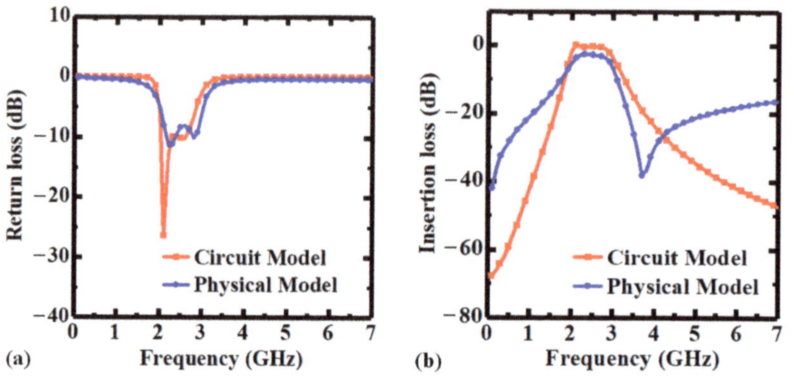

Figure 8. Transmission characteristics of the proposed PI-TSV based BPF: (**a**) return loss (S_{11}), (**b**) insertion loss (S_{21}). The red lines are obtained from the ideal circuit model, and the blue lines are simulated results of the physical model in HFSS.

Table 1 presents the comparisons of the designed PI-TSV based BPF with other published works in respect of RF characteristics, device sizes, technologies, and manufacturing costs and complexities. It is shown that the proposed compact PI-TSV based BPF has comparable RF performance with the state-of-the-art designs, while it occupies a smaller footprint than most of the reported ones. Even though that the filter in [44] has an even smaller size, its fabrication flow is relatively complex as it stacks several parallel plate capacitors on the TGV based inductors. Moreover, as the proposed filter in this work is developed based on a single Si interposer with a thickness of 300 μm and is fully formed by low-cost PI-TSV structures and corresponding RDLs, it is able to be fabricated by the relatively mature TSV technology without the need to introduce extra bonding steps or IC processes. Meanwhile, all TSVs can be synchronously fabricated within one process round, which further reduces the manufacture cost and complexity. Besides, the PI-TSV based BPF is also attractive in the modern RF/electronic micro-systems as it is compatible with 3D integration technology and can be easily integrated with other chips and modules.

Table 1. Comparisons of the proposed BPF with other published designs.

References	f_0 (GHz)	FBW (%)	Insertion Loss (dB)	Return Loss (dB)	Size (mm × mm)	Technology
[17]	2.4	12.5	2.4	15	2.63 × 2.61	LTCC
[18]	2.6	10.2	2.47	20	2 × 1.7	LTCC
[23]	2.4	174	1.1	11	87.5 × 12.8	SSPP
[43]	2.65	62.9	0.6	30	1 × 0.5	TGV and IC
[44]	2.45	31.5	2.4	15	0.44 × 0.33	TGV and IC
[56]	2.7	2	3.8	14	44.8 × 16	DPAM *
[57]	2.33	39.8	0.65	15	18.5 × 18.5	Micro-strip
[58]	2.4	10.4	0.87	15	9.4 × 23.1	Micro-strip
[59]	2.4	12.1	1.2	15	19.1 × 11.2	Micro-strip
This work	2.4	17	2.63	11.4	0.81 × 0.44	PI-TSV

* DPAM: Direct print additive manufacturing.

5. Conclusions

In this work, fully PI-TSV based IPDs including capacitor, inductor, and BPF are designed and evaluated. The impacts of various design parameters on the performance of the TSV based coaxial-like capacitor and the TSV based solenoid inductor are analyzed, providing a guidance to the development of such components. Capacitors and inductors with specific values of capacitances and inductances can be easily achieved by the topologies of the basic coaxial-like capacitor element and solenoid inductor element. Based on this topological method, a novel compact BPF structure is proposed, which is composed of several arrays of coaxial-like capacitors and multi-turns solenoid inductors. The 3-dB bandwidth and the fraction bandwidth (FBW) of the filter is 410 MHz and 17%, respectively. Besides, the in-band insertion loss is less than 2.63 dB, and the return loss in the passband is better than 11.4 dB, showing good RF performance. Notably, the device size of the BPF is only 0.814 mm × 0.444 mm (0.0065 λ_0 × 0.0035 λ_0). Moreover, as the proposed BPF is fully formed by low-cost PI-TSVs with the same structural parameters and corresponding RDLs, it features lower manufacturing cost and complexity compared to other reported designs. Furthermore, the BPF is convenient to be integrated with other devices as it is built in a natural Si interposer structure. As a conclusion, the proposed design of compact BPF is promising for the miniaturization and integration of modern RF applications.

Author Contributions: Conceptualization, Y.D. and Z.Z.; methodology, M.Z. and Z.Z.; software, H.D., H.W., X.P. and M.Z.; validation, M.Z. and Z.Z.; formal analysis, H.D., Y.D., M.Z. and Z.Z.; investigation, H.D., H.W. and X.P.; resources, H.D., Y.D. and Z.Z.; data curation, H.W., X.P., M.Z. and Z.Z.; writing—original draft preparation, Z.Z.; writing—review and editing, H.D., Y.D. and Z.Z.; visualization, M.Z. and Z.Z.; supervision, Y.D. and Z.Z.; project administration, Y.D.; funding acquisition, Y.D. All authors have read and agreed to the published version of the manuscript.

Funding: This research was funded in part by the National Natural Science Foundation of China under Grants 62074015 and 61774015, and in part by the Beijing Nova Program of Science and Technology under Grant Z191100001119078.

Data Availability Statement: Not applicable.

Conflicts of Interest: The authors declare no conflict of interest.

References

1. Pourzadi, A.; Isapour, A.; Kouki, A. Design of Compact Dual-Band LTCC Second-Order Chebyshev Bandpass Filters Using a Direct Synthesis Approach. *IEEE Trans. Microw. Theory Techn.* **2019**, *67*, 1441–1451. [CrossRef]
2. Feng, W.; Gao, X.; Che, W.; Yang, W.; Xue, Q. LTCC Wideband Bandpass Filters with High Performance Using Coupled Lines with Open/Shorted Stubs. *IEEE Trans. Compon. Packag. Manuf. Technol.* **2017**, *7*, 602–609. [CrossRef]
3. Xiao, J.-K.; Zhu, M.; Li, Y.; Tian, L.; Ma, J.-G. High Selective Microstrip Bandpass Filter and Diplexer with Mixed Electromagnetic Coupling. *IEEE Microw. Wirless Compon. Lett.* **2015**, *25*, 781–783. [CrossRef]
4. Zhu, L.; Bu, H.; Wu, K. Aperture Compensation Technique for Innovative Design of Ultra-Broadband Microstrip Bandpass Filter. In Proceedings of the IEEE MTT-S International Microwave Symposium (IMS2000), Boston, MA, USA, 11–16 June 2000; pp. 315–318.
5. Mul, M.; Jasinski, M.; Lamecki, A.; Gómez-García, R.; Mrozowski, M. Inline Microwave Filters with N+1 Transmission Zeros Generated by Frequency-Variant Couplings: Coupling-Matrix-Based Synthesis and Design. *IEEE Trans. Circuits Syst. II Exp. Briefs* **2022**, *69*, 824–828. [CrossRef]
6. Zhang, W.; Gu, J.; Xu, G.; Luo, L.; Li, X. Copper/Benzocyclotene Thin Film Technique Based Microstrip Bandpass Filter Featured by Thick Dielectric Layer for Low Insertion Loss. *Microw. Opt. Technol. Lett.* **2020**, *62*, 3695–3701. [CrossRef]
7. Huang, Y.; Shao, Z.; Liu, L. A Substrate Integrated Waveguide Bandpass Filter Using Novel Defected Ground Structure Shape. *Progr. Electroman. Res.* **2013**, *135*, 201–213. [CrossRef]
8. Sánchez, J.R.; Bachiller, C.; Nova, V.; Boria, V.E. Controlled Out-of-Band Rejection of Filters based on SIW with Alternating Dielectric Line Sections. *IEEE Microw. Wirless Compon. Lett.* **2015**, *29*, 258–260. [CrossRef]
9. Shen, W.; Yin, W.-Y.; Sun, X.-W.; Wu, L.-S. Substrate-Integrated Waveguide Bandpass Filters with Planar Resonators for System-on-Package. *IEEE Trans. Compon. Packag. Manuf. Technol.* **2013**, *3*, 253–261. [CrossRef]
10. Dey, S.; Koul, S.K. Reliable, Compact, and Tunable MEMS Bandpass Filter Using Arrays of Series and Shunt Bridges for 28-GHz 5G Applications. *IEEE Trans. Microw. Theory Techn.* **2021**, *69*, 75–88. [CrossRef]
11. Shim, Y.; Wu, Z.; Rais-Zadeh, M. A High-Performance Continuously Tunable MEMS Bandpass Filter at 1 GHz. *IEEE Trans. Microw. Theory Techn.* **2012**, *60*, 2439–2447. [CrossRef]
12. Mahon, S. The 5G Effect on RF Filter Technologies. *IEEE Trans. Semicond. Manuf.* **2017**, *30*, 494–499. [CrossRef]
13. Lin, W.; Ziolkowski, R.W.; Huang, J. Electrically Small, Low-Profile, Highly Efficient, Huygens Dipole Rectennas for Wirelessly Powering Internet-of-Things Devices. *IEEE Trans. Antennas Propag.* **2019**, *67*, 3670–3679. [CrossRef]
14. Hunter, C.; Billonet, L.; Jarry, B.; Guillon, P. Microwave Filters—Applications and Technology. *IEEE Trans. Microw. Theory Techn.* **2002**, *50*, 794–805. [CrossRef]
15. Liu, L.; Bai, T.-L.; Deng, J.-Y.; Sun, D.; Zhang, Y.; Yong, T.; Zhou, S.-G.; Guo, L.-Z. Substrate Integrated Waveguide Filtering Horn Antenna Facilitated by Embedded Via-Hole Arrays. *IEEE Antennas Wirel. Propag. Lett.* **2020**, *19*, 1187–1191. [CrossRef]

16. Zheng, Y.; Sheng, W. Compact Lumped-Element LTCC Bandpass Filter for Low-Loss VHF-Band Applications. *IEEE Microw. Wirless Compon. Lett.* **2017**, *27*, 1074–1076. [CrossRef]
17. Zhang, X.Y.; Dai, X.; Kao, H.-L.; Wei, B.-H.; Cai, Z.Y.; Xue, Q. Compact LTCC Bandpass Filter with Wide Stopband Using Discriminating Coupling. *IEEE Trans. Compon. Packag. Manuf. Technol.* **2014**, *4*, 656–663. [CrossRef]
18. Xu, J.-X.; Zhang, X.Y.; Zhao, X.-L.; Xue, Q. Synthesis and Implementation of LTCC Bandpass Filter with Harmonic Suppression. *IEEE Trans. Compon. Packag. Manuf. Technol.* **2016**, *6*, 596–604. [CrossRef]
19. Lai, Q.; Fumeaux, C.; Hong, W.; Vahldieck, R. Characterization of the Propagation Properties of the Half-Mode Substrate Integrated Waveguide. *IEEE Trans. Microw. Theory Techn.* **2009**, *57*, 1996–2004.
20. Huang, L.; Cha, H. Compact ridged half-mode substrate integrated waveguide bandpass filter. *IEEE Microw. Wirel. Compon. Lett.* **2015**, *25*, 223–225. [CrossRef]
21. Huang, L.; Cha, H.; Li, Y. Compact Wideband Ridge Half-Mode Substrate Integrated Waveguide Filters. *IEEE Trans. Microw. Theory Techn.* **2016**, *64*, 3568–3579. [CrossRef]
22. Zhao, L.; Zhang, X.; Wang, J.; Yu, W.; Li, J.; Su, H.; Shen, X. A Novel Broadband Band-pass Filter Based on Spoof Surface Plasmon Polaritons. *Sci. Rep.* **2016**, *6*, 36069. [CrossRef]
23. Wang, M.; Sun, S.; Ma, H.F.; Cui, T.J. Supercompact and Ultrawideband Surface Plasmonic Bandpass Filter. *IEEE Trans. Microw. Theory Techn.* **2020**, *68*, 732–740. [CrossRef]
24. Guan, D.-F.; You, P.; Zhang, Q.; Yang, Z.-B.; Liu, H.; Yong, S.-W. Slow-Wave Half-Mode Substrate Integrated Waveguide Using Spoof Surface Plasmon Polariton Structure. *IEEE Trans. Microw. Theory Techn.* **2018**, *66*, 2946–2952. [CrossRef]
25. Guan, D.-F.; You, P.; Zhang, Q.; Xiao, K.; Yong, S.-W. Hybrid Spoof Surface Plasmon Polariton and Substrate Integrated Waveguide Transmission Line and Its Application in Filter. *IEEE Trans. Microw. Theory Techn.* **2017**, *65*, 4925–4932. [CrossRef]
26. Lin, T.-W.; Low, K.K.W.; Gaddi, R.; Rebeiz, G.M. High-Linearity 5.3–7.0 GHz 3-Pole Tunable Bandpass Filter Using Commercial RF MEMS Capacitors. In Proceedings of the 2018 48th European Microwave Conference (EuMC), Madrid, Spain, 25–27 September 2018; pp. 555–558.
27. Hikmat, F.; Ali, M.S.M. RF MEMS Inductors and Their Applications—A Review. *J. Microelectromech. Syst.* **2017**, *26*, 17–44. [CrossRef]
28. Liu, X.; Katehi, L.P.B.; Chappell, W.J.; Peroulis, D. A 3.4–6.2 GHz Continuously Tunable Electrostatic MEMS Resonator with Quality Factor of 460–530. In Proceedings of the 2009 IEEE MTT-S International Microwave Symposium Digest, Boston, MA, USA, 7–12 June 2009; pp. 1149–1152.
29. Hafiz, M.A.A.; Kosuru, L.; Hajjaj, A.Z.; Younis, M.I. Highly Tunable Narrow Bandpass MEMS Filter. *IEEE Trans. Electron. Devices.* **2017**, *64*, 3392–3398. [CrossRef]
30. Koyanagi, M. Heterogeneous 3D Integration—Technology Enabler Toward Future Super-Chip. In Proceedings of the 2013 IEEE International Electron Devices Meeting, Washington, DC, USA, 9–11 December 2013; pp. 1.2.1–1.2.8.
31. Beica, R. 3D Integration: Applications and Market Trends. In Proceedings of the 2015 International 3D Systems Integration Conference (3DIC), Sendai, Japan, 31 August–2 September 2015; pp. TS5.1.1–TS5.1.7.
32. Beyne, E. The Rise of the 3rd Dimension for System Integration. In Proceedings of the 2006 International Interconnect Technology Conference, Burlingame, CA, USA, 5–7 June 2006; pp. 1–5.
33. Wang, Z. 3-D integration and through-silicon vias in MEMS and microsensors. *J. Microelectromech. Syst.* **2015**, *24*, 1211–1244. [CrossRef]
34. Lau, J.H. *Through-Silicon Vias for 3D Integration*; McGraw-Hill: New York, NY, USA, 2012.
35. Motoyoshi, M. Through-Silicon Via (TSV). *Proc. IEEE* **2009**, *97*, 43–48. [CrossRef]
36. Chang, Y.-Y.; Ko, C.-T.; Yu, T.-H.; Hsieh, Y.-S.; Chen, K.-N. Modeling and Characterization of TSV Capacitor and Stable Low-Capacitance Implementation for Wide-I/O Application. *IEEE Trans. Device Mater. Rel.* **2015**, *15*, 129–135. [CrossRef]
37. Tian, U.R.; Yang, R.; Zhuo, C.; Shi, Y. On the Efficacy of Through-Silicon-via Inductors. *IEEE Trans. Very Large Scale Integr. (VLSI) Syst.* **2015**, *23*, 1322–1334.
38. Li, H.; Liu, J.; Xu, T.; Xia, J.; Tan, X.; Tao, Z. Fabrication and Optimization of High Aspect Ratio Through-Silicon-Vias Electroplating for 3D Inductor. *Micromachines* **2018**, *9*, 528. [CrossRef] [PubMed]
39. Kim, J.; Shenoy, R.; Lai, K.-Y.; Kim, J. High-Q 3D RF Solenoid Inductors in Glass. In Proceedings of the 2014 IEEE Radio Frequency Integrated Circuits Symposium, Tampa, FL, USA, 1–3 June 2014; pp. 199–200.
40. Kim, M.S.; Pulugurtha, M.R.; Sundaram, V.; Tummala, R.R.; Yun, H. Ultrathin High-Q 2-D and 3-D RF Inductors in Glass Packages. *IEEE Trans. Compon. Packag. Manuf. Technol.* **2018**, *8*, 643–652. [CrossRef]
41. Yook, J.-M.; Kim, D.; Kim, J.C. High-Q Trenched Spiral Inductors and Low-loss Low Pass Filters Using through Silicon via Processes. *Jpn. J. Appl. Phys.* **2014**, *53*, 04EE11. [CrossRef]
42. Wang, F.; Yu, N. An Ultracompact Butterworth Low-Pass Filter Based on Coaxial Through-Silicon Vias. *IEEE Trans. Very Large Scale Integr. (VLSI) Syst.* **2017**, *25*, 1164–1167. [CrossRef]
43. Sitaraman, S.; Sukumaran, V.; Pulugurtha, M.R.; Wu, Z.; Suzuki, Y.; Kim, Y.; Sundaram, V.; Kim, J.; Tummala, R.R. Miniaturized Bandpass Filters as Ultrathin 3-D IPDs and Embedded Thinfilms in 3-D Glass Modules. *IEEE Trans. Compon. Packag. Manuf. Technol.* **2017**, *7*, 1410–1418. [CrossRef]
44. Qian, L.; Sang, J.; Xia, Y.; Wang, J.; Zhao, P. Investigating on Through Glass via Based RF Passives for 3-D Integration. *IEEE J. Electron. Devices Soc.* **2018**, *6*, 755–759. [CrossRef]

45. Zhang, Z.; Ding, Y.; Chen, Z.; Zhou, M.; Xiao, L.; Cai, Z.; Xiong, M.; Gong, X. Design and Evaluation of a Novel and Ultra-Compact Fully-TGV-Based Self-Shielding Bandpass Filter for 5G Applications. In Proceedings of the 2019 International 3D Systems Integration Conference (3DIC), Sendai, Japan, 8–10 October 2019; pp. 1–4.
46. Mariappan, M.; Fukushima, T.; Beatrix, J.; Hashimoto, H.; Sato, Y.; Lee, K.; Tanaka, T.; Koyanagi, M. Replacing the PECVD-SiO_2 in the Through-Silicon Via of High-Density 3D LSIs with Highly Scalable Low Cost Organic Liner: Merits and Demerits. In Proceedings of the 2014 IEEE 64th Electronic Components and Technology Conference (ECTC), Orlando, FL, USA, 27–30 May 2014; pp. 636–640.
47. Chen, Q.; Cui, H.; Tan, Z.; Wang, Z. Low Capacitance Through-Silicon-Vias with Uniform Benzocyclobutene Insulation Layers. *IEEE Trans. Compon. Packag. Manuf. Technol.* **2013**, *3*, 724–731. [CrossRef]
48. Liu, B.; Yan, Y.; Zhang, Z.; Chen, Z.; Ding, Y. Electrical Characteristics and Thermal Reliability of Blind Through-Silicon-Vias with Polyimide Liners. In Proceedings of the 2016 17th International Conference on Electronic Packaging Technology (ICEPT), Wuhan, China, 16–19 August 2016; pp. 1360–1364.
49. Yan, Y.; Ding, Y.; Chen, Q.; Lee, K.; Fukushima, T.; Koyanagi, M. Vacuum-Assisted-Spin-Coating of Polyimide Liner for High-Aspect-Ratio TSVs Applications. In Proceedings of the 2015 International 3D Systems Integration Conference (3DIC), Sendai, Japan, 31 August–2 September 2015; pp. TS5.2.1–TS5.2.5.
50. Yan, Y.; Zhang, Z.; Cheng, Z.; Zhou, L.; Chen, Z.; Ding, Y. Low Cost Polyimide Liner Formation with Vacuum-Assisted Spin Coating for Through-Silicon-Vias. In Proceedings of the 2016 IEEE International 3D Systems Integration Conference (3DIC), San Francisco, CA, USA, 8–11 November 2016; pp. 1–5.
51. Xiong, M.; Chen, Z.; Ding, Y.; Kino, H.; Fukushima, T.; Tanaka, T. Development of Eccentric Spin Coating of Polymer Liner for Low-Temperature TSV Technology with Ultra-Fine Diameter. *IEEE Electron. Device Lett.* **2019**, *40*, 95–98. [CrossRef]
52. Guiller, O.; Joblot, S.; Lamy, Y.; Farcy, A.; Carpentier, J.F.; Defay, E. Through Silicon Capacitor Co-Integrated with TSVs on Silicon Interposer. *Microelectron. Eng.* **2014**, *20*, 121–126. [CrossRef]
53. Li, J.; Ma, S.; Liu, H.; Guan, Y.; Chen, J.; Jin, Y.; Wang, W.; Hu, L.; He, S. Design, Fabrication and Characterization of TSV Interposer Integrated 3D Capacitor for SIP Applications. In Proceedings of the 2018 IEEE 68th Electronic Components and Technology Conference (ECTC), San Diego, CA, USA, 29 May–1 June 2018; pp. 1974–1980.
54. Lin, Y.; Li, H.Y.; Tan, C.S. Structural Integrity of 3-D Metal-Insulator-Metal Capacitor Embedded in Fully Filled Cu Through-Silicon Via. *IEEE Trans. Compon. Packag. Manuf. Technol.* **2021**, *11*, 918–921. [CrossRef]
55. Lu, K.-C.; Horng, T.-S.; Li, H.-H.; Fan, K.C.; Huang, T.-Y.; Lin, C.-H. Scalable Modeling and Wideband Measurement Techniques for a Signal TSV Surrounded by Multiple Ground TSVs for RF/High-Speed Applications. In Proceedings of the 2012 IEEE 62nd Electronic Components and Technology Conference, San Diego, CA, USA, 29 May–1 June 2012; pp. 1023–1026.
56. Hawatmeh, D.; Weller, T. 2.4 GHz Band Pass Filter Architecture for Direct Print Additive Manufacturing. In Proceedings of the 2018 IEEE/MTT-S International Microwave Symposium—IMS, Philadelphia, PA, USA, 10–15 June 2018; pp. 67–70.
57. Xu, J. Compact Quasi-Elliptic Response Wideband Bandpass Filter with Four Transmission Zeros. *IEEE Microw. Wirless Compon. Lett.* **2015**, *25*, 169–171. [CrossRef]
58. Belmajdoub, A.; El Alami, A.; Das, S.; Madhav, B.T.P.; Bennani, S.D.; Jorio, M. Design, Optimization and Realization of Compact Bandpass Filter Using Two Identical Square Open-Loop Resonators for Wireless Communications Systems. *Int. J. Instrum. (JINST)* **2019**, *14*, 09012. [CrossRef]
59. Afzali, B.; Abbasi, H.; Shama, F.; Dehdasht-Heydari, R. A Microstrip Bandpass Filter with Deep Rejection and Low Insertion Loss for Application at 2.4 GHz Useful Wireless Frequency. *Int. J. Electron. Commun.* **2021**, *138*, 153811. [CrossRef]

micromachines

MDPI

Article

Beam Scanning and Capture of Micro Laser Communication Terminal Based on MEMS Micromirrors

Xuan Wang [1,2,*], Junfeng Han [1,2], Chen Wang [1,2], Meilin Xie [1,2], Peng Liu [1,2], Yu Cao [1,2], Feng Jing [1,2], Fan Wang [1,2], Yunhao Su [1,2] and Xiangsheng Meng [1,2,*]

1 Key Laboratory of Space Precision Measurement Technology, Chinese Academy of Sciences, Xi'an 710119, China
2 Xi'an Institute of Optics and Precision Mechanics, Chinese Academy of Sciences, Xi'an 710119, China
* Correspondence: wangx002@outlook.com (X.W.); mengxiangsheng@opt.ac.cn (X.M.)

Abstract: With the development of space laser communication and the planned deployment of satellite Internet constellations, there is a growing demand for microminiature laser communication terminals. To meet the requirements of size, weight and power (SWaP), miniaturized terminals require smaller drive components to complete on-orbit scanning and capture, which must be fast and efficient to enable satellite laser communication networks. These miniaturized laser communication terminals are highly susceptible to the impact of the initial pointing accuracy of the laser beam and microvibrations of the satellite platform. Therefore, this paper proposes a laser scanning-capture model based on a Micro-electromechanical Systems (MEMS) micromirror that can provide a fast, large-scale scanning analysis. A scanning overlap factor is introduced to improve the capture probability under the influence of microvibrations. Finally, experimental analysis was carried out to verify the effectiveness of the proposed model, which can establish a theoretical basis for future ultra-long-distance microspace laser communication.

Keywords: free-space optical communication; MEMS micromirror; laser scanning and capture

Citation: Wang, X.; Han, J.; Wang, C.; Xie, M.; Liu, P.; Cao, Y.; Jing, F.; Wang, F.; Su, Y.; Meng, X. Beam Scanning and Capture of Micro Laser Communication Terminal Based on MEMS Micromirrors. *Micromachines* **2023**, *14*, 1317. https://doi.org/10.3390/mi14071317

Academic Editor: Stefan Wabnitz

Received: 19 May 2023
Revised: 16 June 2023
Accepted: 24 June 2023
Published: 27 June 2023

1. Introduction

Compared to traditional RF technology, free-space optical communication offers significant advantages in data rate performance, cost-effectiveness, and increased security [1]. As such, it plays an important role in space-ground integrated networks. At the same time, CubeSat satellites are becoming increasingly popular in low Earth orbit networks due to their low cost, quick response times, potential for constellation formation, and ability to perform tasks that a single large satellite cannot [2,3]. Since the German Aerospace Center began to focus on the research and development of small satellite optical communication terminals more than 10 years ago, this field has attracted more and more attention from major research institutions: NASA, the Jet Propulsion Laboratory, European Space Agency, and National Institute of Information and Communication Technology of Japan [4,5].

In 2008, the German TerraSAR-X and the United States NFIRE satellites marked a new phase in space laser communication by successfully establishing the first coherent link over a distance of more than 5000 km at a rate of 5.6 Gbit/s [6,7]. The TerraSAR-X satellite is equipped with the LCT135 laser terminal developed by DLR and has become the standard configuration of the European Data Relay Satellite System (EDRS). LCT135 has made significant contributions to the establishment of high-orbit backbone satellite networks and low-orbit satellite constellations. After more than 30 years of research and technological accumulation, TESAT has developed a range of new laser communication terminal (LCT) products [8,9].

However, establishing optical communication links between CubeSats is a daunting task. The laser transmission link requires very strict beam pointing stability because the system is extremely susceptible to interference from the mechanical vibration of the loading platform. Severe jitter will cause the laser beam to deviate from the field of view of the target receiving system, resulting in pointing errors that increase the probability of signal interruption and information loss. Precise pointing, acquisition, and tracking (PAT) of incident light to establish a communication link often requires complex optical and hardware systems to account for the relative movement between terminals. For small satellites with strict size, weight and power (SWaP) requirements, the PAT system also increases the system burden. Therefore, it is vital that the miniaturized laser communication precise pointing transceiver system be studied.

Mirror-based PAT systems use reflective mirrors as actuators to achieve beam stabilization, pointing, and tracking. The reflector mirrors deflect the incident laser beam onto the receiving sensor, but unlike traditional gimbal-based PAT mechanisms, mirror-based PAT systems are lightweight, have high response speeds, and provide fine pointing resolution. Feedback control systems are typically used to detect errors in the trajectory or phase of the reflected beam and to fine-tune the deflection angle of the reflector mirror.

Micro-electromechanical Systems (MEMS) is a cutting-edge technology based on micro/nanotechnology [10]. It enables various micro-actuators, sensors, and signal processing circuits to be integrated into a microscale/nanoscale chip. Because of its advantages—small size, ultra-low power consumption, low cost, high integration density, and stable reliability—MEMS is increasingly important for the entire semiconductor ecosystem. It has been widely used in defense, high-speed optical communication, the aerospace and automotive industries, medical equipment, and smart terminal devices [11–13].

2. MEMS Micromirror Optical Scanning Model

To investigate the relationship between the deflection angle and planar scanning of the 2D MEMS micromirror, a spatial rectangular coordinate system was established (Figure 1). In the initial state, where it did not deflect, the micromirror coincided with the XOY plane, and $\vec{N}$ was the normal unit vector of the XOY plane coincident with the Z axis. The incident light $\vec{A}$ formed an angle β with the normal of the MEMS reflection plane on the XOZ plane. The reflected light $\vec{A}'$ was reflected by the MEMS reflective plane, entered the $X'O'Z'$ plane, and was coplanar with it. The distance between O and O' was L, the unit normal vector was $\vec{N}$, and the output light was $\vec{A}'$. The surface of the MEMS reflective mirror continuously changes as the 2D MEMS mirror vibrates in the horizontal and vertical directions, and this relationship can be derived using the method of matrix optics [12,14].

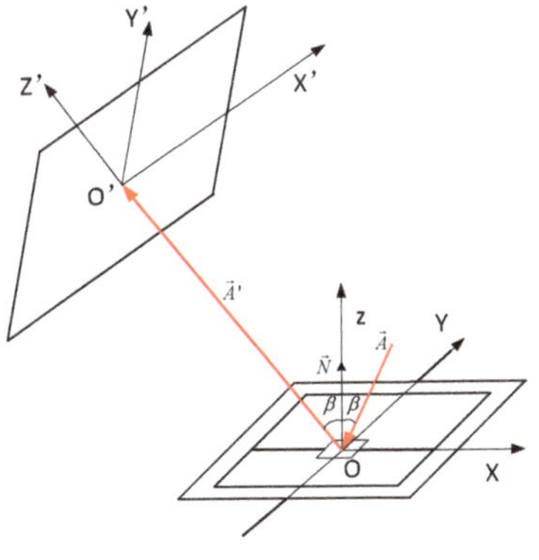

Figure 1. Optical scanning model of 2D MEMS micromirror.

The vector form of its reflection law can be expressed in matrix optics by Equation (1).

$$\vec{A}' = \vec{A} - 2(\vec{A} \cdot \vec{N}) \cdot \vec{N} \tag{1}$$

where $\vec{A}$ is the incident light unit vector; $\vec{N}$ is the normal direction unit vector; and $\vec{A'}$ is the outgoing light unit vector. Transforming the vector form of the reflection law in Equation (1) into a matrix expression, establishes a 3D rectangular coordinate system x, y, z where i, j, k represent the directions of the three-dimensional components in the vector analysis, respectively.

$$\vec{A} = A_x\vec{i} + A_y\vec{j} + A_z\vec{k}. \tag{2}$$

$$\vec{A'} = A'_x\vec{i} + A'_y\vec{j} + A'_z\vec{k}. \tag{3}$$

$$\vec{N} = N_x\vec{i} + N_y\vec{j} + N_z\vec{k}. \tag{4}$$

The outgoing light $\vec{A'}$, the incident light A and the reflection matrix V can be expressed as Equation (5).

$$A' = AV. \tag{5}$$

According to the above description, the expression of the reflection matrix can be obtained as Equation (6).

$$V = \begin{bmatrix} 1-2N_x^2 & -2N_yN_x & -2N_zN_x \\ -2N_xN_y & 1-2N_y^2 & -2N_zN_y \\ -2N_xN_z & -2N_yN_z & 1-2N_z^2 \end{bmatrix}. \tag{6}$$

The expression of the reflection matrix V is only related to the component of the normal unit vector $\vec{N}$ of the reflection surface in each direction, and the reflection matrix is determined by the spatial position of the reflection surface. When the MEMS micromirror did not vibrate in the initial state using $X'Y'Z'$ as the principal coordinate system, the following relationship was obtained:

$$\vec{A} = \begin{bmatrix} A_x \\ A_y \\ A_z \end{bmatrix} = \begin{bmatrix} -\sin 2\beta \\ 0 \\ -\cos 2\beta \end{bmatrix}. \tag{7}$$

$$\vec{N} = \begin{bmatrix} N_x \\ N_y \\ N_z \end{bmatrix} = \begin{bmatrix} -\sin \beta \\ 0 \\ -\cos \beta \end{bmatrix}. \tag{8}$$

Thus, the reflection matrix in the initial state can be obtained as Equation (9).

$$A' = AV = \begin{bmatrix} -\sin 2\beta & 0 & -\sin\beta \\ 0 & 1 & 0 \\ -\cos 2\beta & 0 & -\cos\beta \end{bmatrix} \begin{bmatrix} -\sin 2\beta \\ 0 \\ -\cos 2\beta \end{bmatrix} = \begin{bmatrix} 0 \\ 0 \\ 1 \end{bmatrix}. \tag{9}$$

Equation (9) confirms that the intersection point of $\vec{A'}$ and the $X'O'Y'$ plane in the initial state is the origin O', and the outgoing light overlaps with the Z' axis. When the MEMS mirror undergoes vibrations and deflections in the 2D space, the aforementioned theory will be used to calculate the intersection coordinates of the projection line and the $X'O'Y'$ plane as well as the deflection angle of the Z' axis. Assuming that the horizontal deflection angle of the MEMS micromirror is α_x, the vertical deflection angle is α_y. This is equivalent to the XOY plane in Figure 1: first rotating α_y counterclockwise with the y-axis as the axis of rotation and then rotating α_x clockwise with the x-axis as the axis of rotation. The new normal unit vector corresponding to the reflective surface of the MEMS micromirror can be expressed as Equation (10).

$$\vec{N} = \begin{bmatrix} N_x \\ N_y \\ N_z \end{bmatrix} = \begin{bmatrix} \sin(\beta + \alpha_x)\cos\alpha_y \\ \sin\alpha_y \\ \cos(\beta + \alpha_x)\cos\alpha_y \end{bmatrix}. \tag{10}$$

After calculation and simplification, the output light in the deflected state of the mirror was obtained:

$$A' = AV = \begin{bmatrix} \cos^2\alpha_y \sin 2\alpha_x - \sin^2\alpha_y \sin 2\beta \\ \cos(\beta - \alpha_x)\sin 2\alpha_y \\ \cos^2\alpha_y \cos 2\alpha_x - \sin^2\alpha_y \sin 2\beta \end{bmatrix}. \tag{11}$$

Regardless of the deviation of the reflecting mirror, the starting point of the output light was always $O'(0,0,-L)$. Therefore, the linear equation of the output light was

$$\frac{X'}{A'_x} = \frac{Y'}{A'_y} = \frac{Z' + L}{A'_z}. \tag{12}$$

The coordinates of the intersection point between the incident light and the $X'O'Y'$ plane was obtained as follows:

$$\begin{cases} X' = L\frac{\cos^2\alpha_y \sin 2\alpha_x - \sin^2\alpha_y \sin 2\beta}{\cos^2\alpha_y \cos 2\alpha_x - \sin^2\alpha_y \cos 2\beta} \\ Y' = L\frac{\cos(\beta-\alpha_x)\sin 2\alpha_y}{\cos^2\alpha_y \cos 2\alpha_x - \sin^2\alpha_y \cos 2\beta} \end{cases}. \tag{13}$$

It can be seen from Equation (13) that when $\alpha_y = 0$, the MEMS micromirror was deflected only in the horizontal direction, and $X = L\tan 2\alpha_x$ was obtained, indicating that the new reflected line was deflected at an angle of $2\alpha_x$ to the original reflected line. This meant that the optical reflection angle in the horizontal direction was twice the mechanical deflection angle of the micromirror. On the other hand, when $\alpha_x = 0$ the relationship was not satisfied in the vertical direction. Since the condition for this conclusion is that the initial incident light was horizontally incident, it lay in the horizontal plane. When $\alpha_x \to 0, \alpha_y \to 0$, indicating that the deviation angle of the MEMS mirror was extremely small, we approximated $\sin\alpha \to 0, \cos\alpha \to 1$, which implied $Y/X \to \cos\beta$. The equation showed that when the mechanical deflection angle was small, the ratio of the vertical scan length to the horizontal scan length of the reflector was approximately the cosine of the incident light angle; that is, the scanning effect of the MEMS micromirror was related to the incident angle of the initial incident light.

From Equation (13), it is clear that for various horizontal mechanical deflection angles α_x and vertical mechanical deflection angles α_y of the MEMS micromirrors as well as the different initial light incident angles, the relationship between the horizontal and the vertical deflection angles of the initial output light vector could be expressed as Equation (14):

$$\begin{cases} \varphi_x = \tan^{-1}\frac{X}{L} = \frac{\cos^2\alpha_y \sin 2\alpha_x - \sin^2\alpha_y \sin 2\beta}{\cos^2\alpha_y \cos 2\alpha_x - \sin^2\alpha_y \cos 2\beta} \\ \varphi_y = \tan^{-1}\frac{Y}{L} = \frac{\cos(\beta-\alpha_x)\sin 2\alpha_y}{\cos^2\alpha_y \cos 2\alpha_x - \sin^2\alpha_y \cos 2\beta} \end{cases}. \tag{14}$$

3. Scanning Range of MEMS Micromirrors

As scanning devices in laser communication, MEMS micromirrors have the significant advantages of being small and giving excellent performance [15]. However, their mechanical deflection angle is relatively limited, and currently available rapid-scanning 2D scanning mirrors typically have mechanical deflection angles of around $\pm 5°$, rendering them unsuitable for large-field scanning. In view of this, the expansion angle optical system for the MEMS micromirror is designed to increase the scanning angle. By adjusting the distance between the lens and the micromirror, the optical system expands or reduces the optical scanning range [16].

By designing an optical beam expanding system as shown in Figure 2, the laser beam is emitted and then collimated before being focused onto the reflective surface of a MEMS mirror by a preceding focusing lens. The mechanical deflection angle of the MEMS mirror is relatively small. After passing through an $f-\theta$ lens, a light spot corresponding to a certain field of view for the rear lens is obtained. In this case, the field-of-view angle is designed to be 60°.

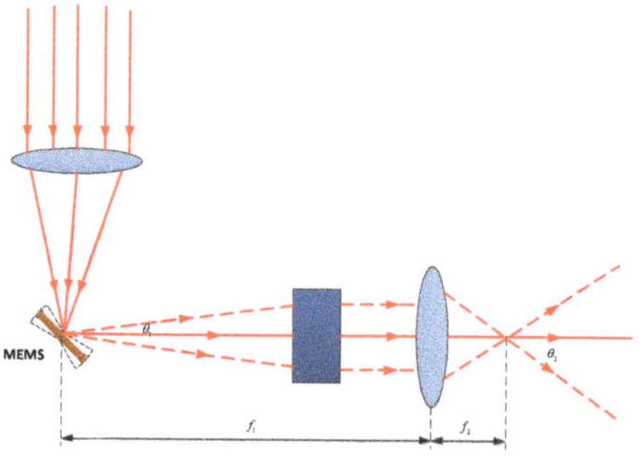

Figure 2. Schematic diagram of the principle of the MEMS micromirror scanning-expansion angle.

The maximum image height y on the image plane of the $f-\theta$ lens can be expressed as:

$$y = f_1 \cdot \theta_1 \tag{15}$$

where f_1 is the focal length of the $f-\theta$ lens, and θ_1 is 2 times the maximum mechanical deflection angle of the MEMS. Finally, through a positive lens, its field of view θ_2 satisfies Equation (16):

$$y = f_2 \cdot \tan\theta_2 \tag{16}$$

where f_2 is the effective image square focal length of the angle-expanding lens, and θ_2 is the field-of-view angle after beam expansion. Therefore, Equation (17) can be obtained:

$$\theta_2 = \arctan\left(\frac{f_1}{f_2} \cdot \theta_1\right). \tag{17}$$

Obviously, to expand the scanning-angle range of the MEMS micromirror, f_2 must be smaller than f_1, which is a key condition of the optical beam expander system. Moreover, when the light emitted by the short-focus positive lens is parallel or close to parallel beams, the laser spot can be made smaller when projected from a longer distance.

4. MEMS Micromirror Scanning Capture System

The microsized laser communication terminal scanning system based on MEMS micromirrors is illustrated in Figure 3. The beam emitted by the laser diode is reflected by the MEMS micromirror and then passes through an optical telescope for emission. The optical system magnifies the beam scanning angle. Similarly, the received light is focused onto a photodetector through an optical coupling lens. The detected beam deflection error is fed back to the micromirror controller for real-time correction of the beam deflection angle.

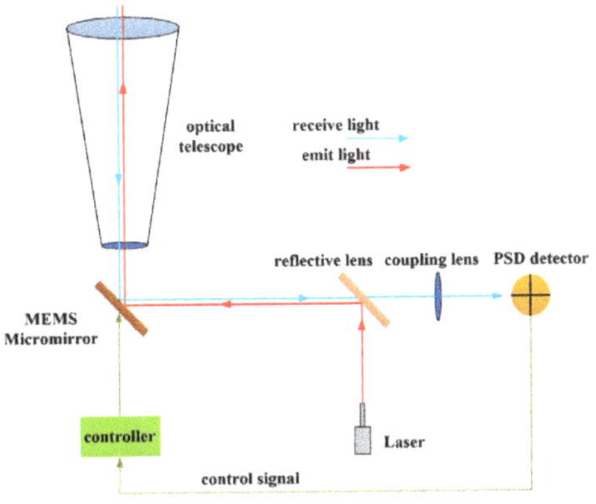

Figure 3. Schematic diagram of laser communication capture and scanning system based on MEMS.

4.1. MEMS Micromirror Scanning Mode

Based on the different modes of mechanical vibration, MEMS micromirrors can be classified into quasistatic and resonant types. The former operates at non-resonant frequencies and can be locked at a certain angle or perform vector graphics. The latter operates at resonant frequencies, allowing for larger scanning angles in resonant mode. Depending on the application requirements and fabrication capabilities, both quasistatic and resonant designs can be chosen when designing MEMS 2D scanning micromirrors: dual-axis resonant, dual-axis quasistatic, and resonant with quasistatic. The dual-axis resonant configuration generates Lissajous scanning patterns, but it places higher demands on the resonant frequency compared to the other two at the same spatial resolution. Combining a resonant axis with a quasistatic axis overcomes this drawback and enables grating scanning patterns with larger scanning angles for the resonant axis compared to dual-axis quasistatic scanning. The dual-axis quasistatic configuration is better suited for vector displays and beam control applications.

The three operating modes are primarily designed to address different operational scenarios and requirements. Various applications such as laser projection, LiDAR, and 3D imaging demand fast scanning capabilities without the need for static operation. On the other hand, laser communication requires precise beam stabilization during capture, alignment, and the establishment of the communication link. This requires MEMS micromirrors to operate in a static mode and perform closed-loop tracking based on position feedback. Therefore, in laser communication, MEMS microreflectors work in quasistatic mode for both axes of deflection.

Currently, commonly used methods for laser communication include spiral, raster, and raster–spiral scanning. In this paper, we propose equidistant and equilinear velocity scanning because rectangular raster scanning can cover uncertain areas effectively and is relatively easy to implement. However, it does not prioritize scanning from positions that have higher target–appearance probability. Instead, it treats the entire uncertain area equally, resulting in lower capture efficiency. Moreover, for circular uncertain areas, rectangular-raster and rectangular-spiral scanning have the disadvantage of increasing additional scanning time around the periphery. Spiral scanning is a traditional method that starts from the center point and spirals outward. Its advantage is that it covers uncertain areas quickly, but because the angular velocity remains constant during scanning, the disadvantage is that the linear velocity increases. As the number of spiral cycles increases so does the distance between adjacent scanning points, which leads to missed scans. On the other hand, equidistant and equilinear velocity spiral scanning overcomes this limitation by transforming a constant angular-velocity scan into a spiral scan that has equidistant and equilinear velocity, which improves the beam coverage of uncertain areas. Starting from the center point with the highest target appearance probability, this method scans uncertain area based on the trajectory of the spiral line, resulting in a higher capture probability. Moreover, the scanning pattern conforms to the circular shape of the uncertain area. A schematic diagram of equidistant and equilinear velocity spiral scanning is shown in Figure 4.

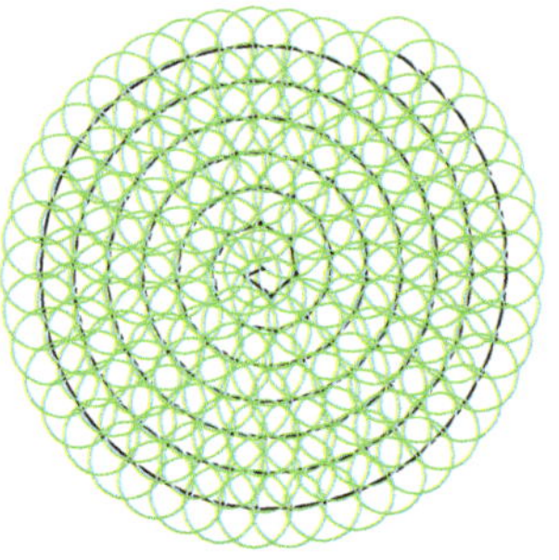

Figure 4. The schematic diagram of equidistant and equilinear velocity spiral scanning.

4.2. MEMS Micromirror Scan Time

In polar and Cartesian coordinate systems, the equidistant and equilinear velocity spiral curve can be represented as follows:

$$\begin{cases} \rho = \frac{I_\theta}{2\pi}\theta \\ I_\theta = (\frac{\sqrt{2}}{2} - k_0)\theta_b \end{cases} \tag{18}$$

$$\begin{cases} x = \frac{I_\theta}{2\pi}\theta\cos(\theta) \\ y = \frac{I_\theta}{2\pi}\theta\sin(\theta) \end{cases} \tag{19}$$

where θ_b represents the divergence angle of the laser beam; I_θ represents the spacing between adjacent spiral curves, also known as the scanning step size, which is the distance between two scanning points; k_0 is the additional overlap factor; ρ represents the polar radius of the scanning spiral curve; and θ represents the polar angle of the scanning spiral curve. Let the time interval between adjacent scanning points be Δt, which represents the dwell time of the beam at a specific point during scanning. In the absence of beacons, the dwell time is only related to the bandwidth F_{ac} of the actuator, denoted as $\Delta t = \frac{1}{F_{ac}}$. Therefore, the spiral scanning speed can be represented as Equation (20):

$$v_0 = \frac{I_\theta}{\Delta t}. \tag{20}$$

In conclusion, the expression of the length l of the spiral scanning line in the polar coordinate system can be obtained by the following equations.

$$l = \frac{I_\theta}{4\pi}[\theta\sqrt{1+\theta^2} + \ln(\theta + \sqrt{1+\theta^2})]. \tag{21}$$

When the polar angle θ of the scanning spiral curve is greater than π, the length l of the spiral scanning line can be simplified to

$$l \approx \frac{\pi\rho^2}{I_\theta} = \frac{I_\theta}{4\pi}\theta^2. \tag{22}$$

Therefore, when the polar angle of the scanning spiral curve is θ, the scanning time t is

$$t = \frac{\Delta t}{4\pi}\theta^2. \tag{23}$$

When the scanning polar radius ρ is the half angle θ_u of the capture uncertainty zone, $\rho = \theta_u$. After scanning the entire capture uncertain area, the spiral curve reaches its maximum polar angle $\theta_{\max}$ as follows:

$$\theta_{\max} = \frac{2\pi}{I_\theta}\theta_u. \tag{24}$$

The scanning time t_u required to scan the entire capture uncertainty region is given as

$$t_u = \frac{\pi\theta_u^2}{I_\theta^2}\Delta t. \tag{25}$$

For a $\pm 30^\circ$ uncertain area with a scanning accuracy of 0.1°, a laser beam divergence angle of 1° and a MEMS micromirror scanning bandwidth of 50 Hz, the scanning time can be obtained through simulation, as shown in Figure 5.

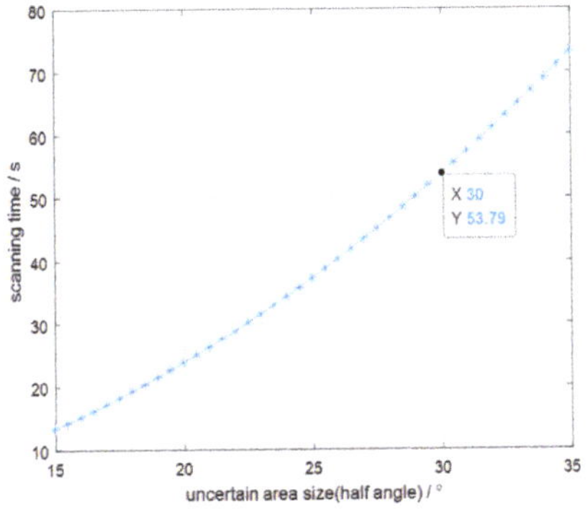

Figure 5. The relationship between the laser beam scanning time and the size of the uncertain region.

From the results, it can be seen that as the half-angle of the uncertain area gradually increases, so does the scanning time. When the area half-angle is 30°, the time is 53.8 s.

5. Scan Capture Simulation Experiment

Due to the small divergence angle of the communication light that MEMS micromirrors use for laser communication scanning and capturing, the beam jitter impact is significant. Moreover, the system operates in an open-loop state during scanning, without error correction, which can cause the scanning beam to deviate from its intended trajectory, leading to excessive overlap or missed scanning spots. Ultimately, this results in decreased coverage of the scanning beam over the uncertain area.

Line-of-sight (LOS) jitter is caused mainly by the vibration of the platform. Therefore, this paper proposes mitigating the effect of optical axis jitter by introducing an additional overlap factor during scanning. In general, LOS jitter follows a Rayleigh distribution in the radial direction, and its probability density function can be expressed as

$$f(\alpha) = \frac{\alpha}{\varphi^2}\exp(-\frac{\alpha^2}{2\varphi^2}). \tag{26}$$

During scanning, α is the angular deviation of the scanning-beam jitter φ is the standard deviation of the beam. When the angular deviation α of the jitter exceeds a certain threshold in the azimuth and elevation directions, there is a gap between the scanning spots, resulting in an uncovered scanning area. This phenomenon is known as "missed scanning". Let's designate the threshold angle for missed scanning as γ and the probability of missed scanning as P_l. The expression for P_l is

$$P_l = \int_{\gamma}^{\infty} f(\alpha) d\alpha = \int_{\gamma}^{\infty} \frac{\alpha}{\varphi^2} \exp(-\frac{\alpha^2}{2\varphi^2}) d\alpha = \exp(-\frac{\gamma^2}{2\varphi^2}). \tag{27}$$

Therefore, the coverage P_s of the scanning beam of the communication light on the capture uncertainty region is shown by Equation (28).

$$P_s = 1 - P_l = 1 - \exp(-\frac{\gamma^2}{2\varphi^2}). \tag{28}$$

From this, the expression of the missed scan-limit angle γ can be deduced as

$$\gamma = \varphi \sqrt{-2\ln(1 - P_s)}. \tag{29}$$

To eliminate the missed scan and meet the scanning beam coverage requirements P_s of the uncertain region, an additional overlap factor k_0 was introduced, as shown in Equation (30):

$$k_0 = \frac{2\gamma}{\theta_b} = \frac{2\varphi\sqrt{-2\ln(1 - P_s)}}{\theta_b}. \tag{30}$$

To further analyze and validate the impact of factors such as beam divergence angle, overlap factor, size of the uncertain region and possible microvibrations on the performance of laser communication scanning capture, experimental studies were conducted using computer simulations. For this purpose, a GUI experimental interface was designed as shown in Figure 6. This software program allows the flexible adjustment of parameters such as beam divergence angle, overlap factor, detector response time, size of the uncertain region, and standard deviation of microvibrations. By changing the values of these parameters, the scanning trajectory of the spot and its coverage of the uncertain region can be observed in the interface, enabling simulation-based calculations of the time required to complete the scanning of the uncertain region.

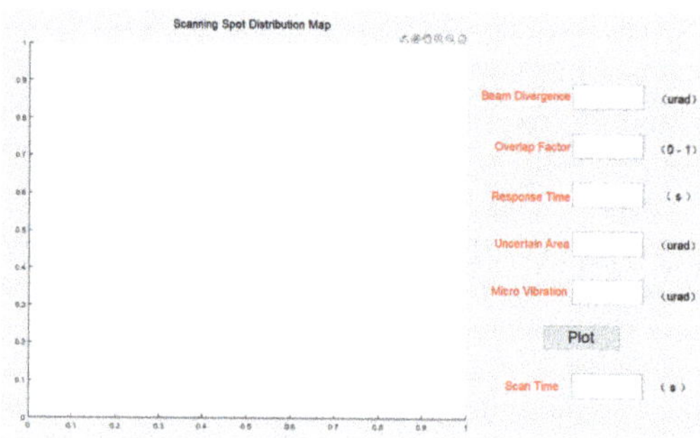

Figure 6. Laser beam scanning capture simulation experiment interface.

First, the simulation program analyzed the scanning coverage of the same uncertain region for different beam divergence angles. The results, shown in Figure 7, indicated that when the system was affected by microvibrations to some extent, certain areas exhibited missed scanning. This observation was consistent with the previous analysis. As the beam divergence angle increased, the time required to scan the entire uncertain region gradually decreased. It was observed that while the microvibrations were kept constant, a larger beam divergence angle led to a smaller area of missed scanning, suggesting that increasing the beam divergence angle had a certain effect on reducing the impact of microvibrations.

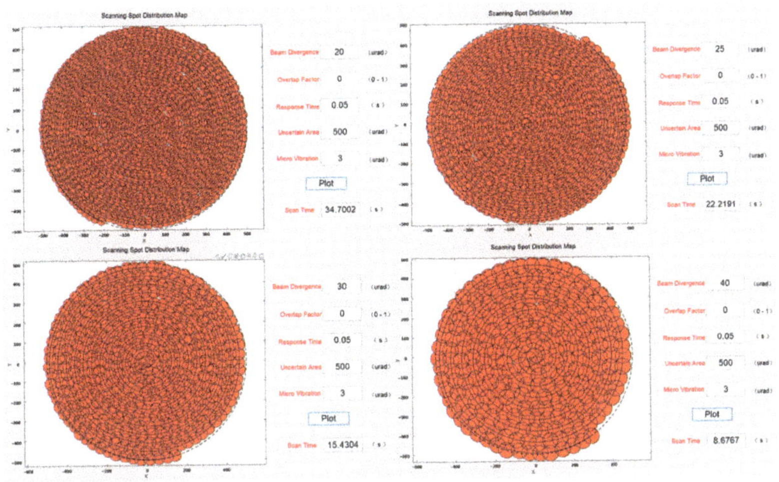

Figure 7. Scanning capture spot distribution at different beam divergence angles.

Next, while keeping the beam divergence angle constant, the scanning coverage for different sizes of uncertain regions was analyzed. The simulation results, shown in Figure 8, revealed that when the system was affected by microvibrations, changing the size of the uncertain region without altering the beam divergence angle and coverage factor did not reduce missed scans. Moreover, under the same parameter settings, larger uncertain regions required a longer time to complete the scanning. Furthermore, to mitigate missed scanning at the same beam divergence angle, it was necessary to increase the coverage factor, which further increased the scanning duration.

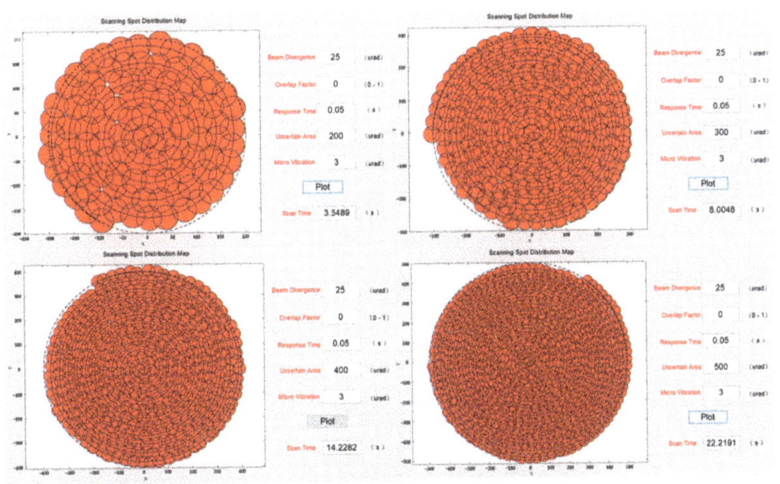

Figure 8. Scanning capture spot distribution at different uncertain regions.

To verify the reduction in missed scanning caused by microvibrations, the overlap factor was gradually increased from 0 to 0.4 while keeping other system parameters constant. The simulation results, shown in Figure 9, indicated that when the system was affected by microvibrations, severe missed scanning occurred if no overlap factor had been introduced. However, when it was introduced and gradually increased to 0.3, there was a noticeable reduction in microvibrations with a standard deviation of 5 μrad. Until the overlap factor reached 0.4, there were no missed scans. However, it is important to note that as the overlap factor increased, scanning time also increased, which should not be ignored. At an overlap factor of 0.4, the time required for the system to scan the entire uncertain region increased by 68% compared to no overlap factor. In practical applications, it is necessary to consider both the scanning time and coverage rate to select optimal parameters.

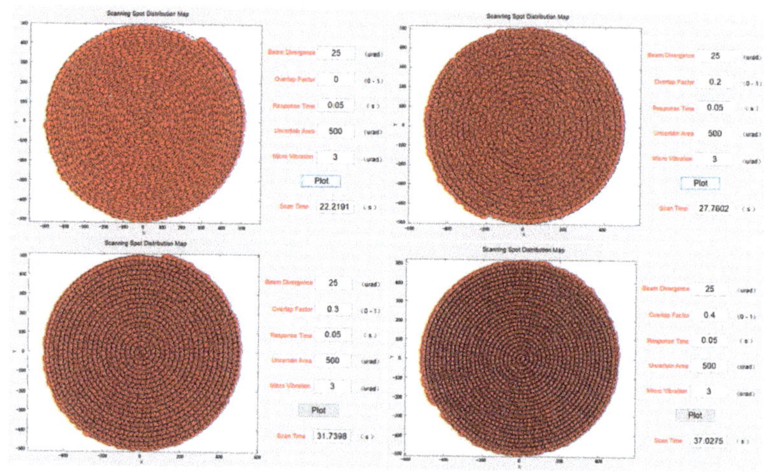

Figure 9. Scanning capture spot distribution at different stacking factors.

Finally, to verify the effect of microvibrations on the system visually, standard deviations of 3, 5, 10, and 20 μrad were introduced under the same beam divergence angle, overlap factor, and uncertain region size. The scanning simulation results in Figure 10 revealed that as the effect of microvibrations gradually increased, the occurrences of missed scanning became more severe.

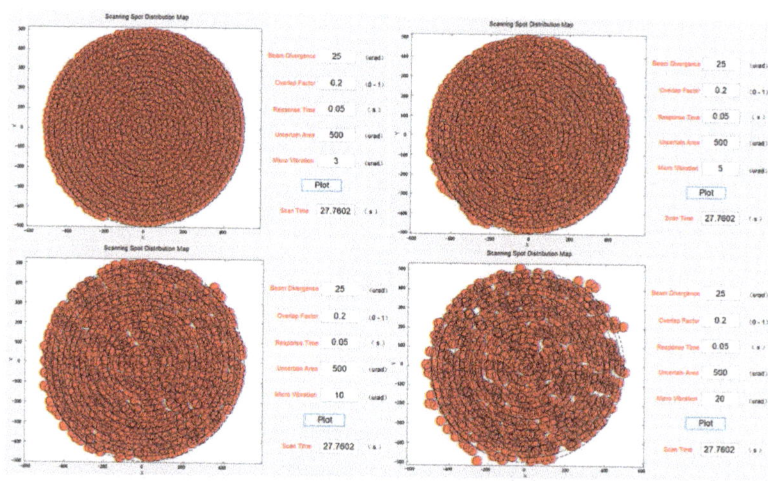

Figure 10. Scanning capture spot distribution under the influence of different microvibrations.

As mentioned earlier, to mitigate the effect of microvibrations, increasing the size of the overlap factor was considered. Building upon the previous set of simulation experiments, the overlap factor was gradually increased up to 0.5 while the microvibration standard deviation remained at 20 μrad. The scanning simulation results at this stage are shown in Figure 11. With an overlap factor of 0.4, there were still some instances of missed scanning, but the beams had already covered the majority of the uncertain region. When the overlap factor reached 0.5, there were no missed scans. However, the scanning time of the system reached 44 s, representing a 95% increase compared to the 0.2 overlap factor.

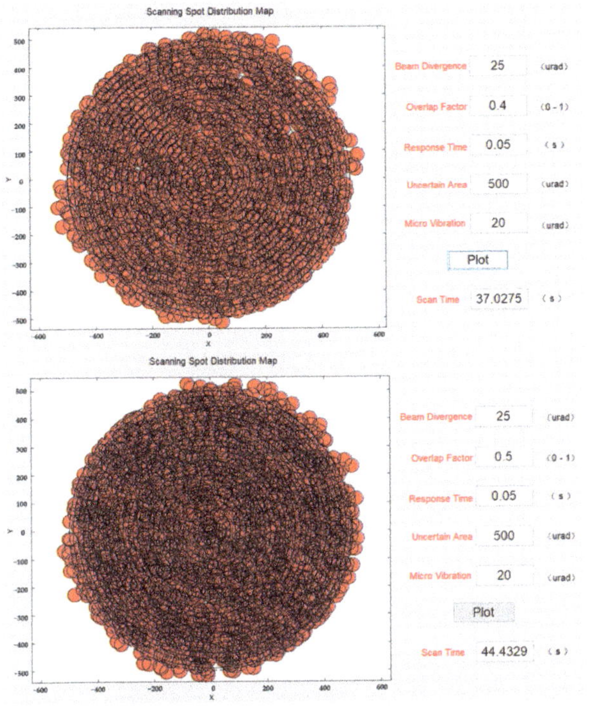

Figure 11. Reduction in missing scans by increasing the overlap factor.

Based on the above analysis, it is evident that the system inevitably experiences some level of micro-vibration effects. These micro-vibrations introduce missed scanning occurrences during the scanning process, which adversely impact the scanning capture probability. To overcome the detrimental effects of micro-vibrations in equidistant and equilinear velocity spiral scanning, the introduction of an overlap factor is necessary. However, increasing the overlap factor also increases the system scanning time. Therefore, in practical applications, a comprehensive consideration of the overlap factor, balancing capture probability and scanning time, is required. Through simulation analysis, the optimal system parameters can be determined.

6. Conclusions

This paper presented a laser beam scanning and capture model based on MEMS micromirrors that focused on fast and wide-range scanning strategies. By introducing a scanning overlap factor, the capture probability under the influence of microvibrations was enhanced. Experimental analysis validated the effectiveness of the scanning capture strategy because platform microvibration was one of the main factors affecting laser communication scanning and capture. When the system experienced certain levels of microvibration, missed scans occurred. As the beam divergence angle gradually increased, the time required to scan the entire uncertain region decreased. Increasing the beam divergence angle while keeping microvibrations constant led to smaller missed scanning areas, indicating the effectiveness of increasing the beam divergence angle to reduce microvibrations. Changing the size of the uncertain region without altering the beam divergence angle or overlap factor did not improve missed scanning caused by microvibrations. Moreover, under the same parameter settings, larger uncertain regions required more time to complete scanning. If mitigating missed scanning occurrences with a constant beam divergence angle were desired, increasing the overlap factor became necessary, but this further increased the scanning time. In practical applications, a comprehensive consideration of scanning time and coverage rate is required to select optimal parameters. The scanning capture model established in this paper provides a theoretical foundation and technical support for practical applications.

Author Contributions: Conceptualization, X.W., J.H. and X.M.; methodology, X.W. and X.M.; software, X.W. and C.W.; validation, P.L., Y.C. and F.J.; formal analysis, X.W. and F.W.; investigation, X.W.; resources, X.W. and Y.S.; data curation, X.W.; writing—original draft preparation, X.W.; writing—review and editing, X.W. and X.M.; visualization, X.W.; supervision, J.H. and M.X.; project administration, J.H. All authors have read and agreed to the published version of the manuscript.

Funding: This research was funded by The National Key R&D Program of China, grant number 2022YFC2203800.

Data Availability Statement: The data will be made available at a reasonable request to the corresponding author.

Acknowledgments: The authors express sincere thanks for the experiments provided by the Photoelectric Tracking and Measurement Technology Laboratory, Xi'an Institute of Optics and Precision Mechanics, CAS.

Conflicts of Interest: The authors declare no conflict of interest.

References

1. Heine, F.; Sánchez-Tercero, A.; Martin-Pimentel, P.; Höpcke, N.; Hasler, D.; Zech, H. In orbit perfomance of tesat LCTs. In Proceedings of the Free-Space Laser Communications XXXI, San Francisco, CA, USA, 4–6 February 2019; SPIE: Bellingham, WA, USA, 2019; Volume 10910, pp. 181–188.
2. Ishola, F.; Carrasco-Casado, A.; Cordova, R.; Masui, H.; Munemasa, Y.; Trinh, P.V.; Fuse, T.; Tsuji, H.; Cho, M.; Toyoshima, M. Characterization and Comparison of CubeSat and Drone Platform Jitter Effects on Laser Beam Pointing Stability. *Prepr. Org.* **2022**, *2022*, 080206. [CrossRef]
3. Liang, Y.; Su, X.; Cai, C.; Wang, L.; Liu, J.; Wang, H.; Wang, J. Adaptive turbulence compensation and fast auto-alignment link for free-space optical communications. *Opt. Express* **2021**, *29*, 40514–40523. [CrossRef]
4. Yu, S.; Ma, Z.; Wu, F.; Ma, J.; Tian, L. Overview and trend of steady tracking in free-space optical communication links. In *Selected Papers from Conferences of the Photoelectronic Technology Committee of the Chinese Society of Astronautics 2014, Part I*; SPIE: Bellingham, WA, USA, 2015; Volume 9521, pp. 141–150.
5. Yao, H.; Wang, L.; Wang, X.; Lu, Z.; Liu, Y. The space-terrestrial integrated network: An overview. *IEEE Commun. Mag.* **2018**, *56*, 178–185. [CrossRef]
6. Israel, D.J.; Shaw, H. Next-generation NASA Earth-orbiting relay satellites: Fusing optical and microwave communications. In Proceedings of the 2018 IEEE Aerospace Conference, Big Sky, MT, USA, 3–10 March 2018; IEEE: Piscataway, NJ, USA, 2018; pp. 1–7.
7. Li, L.; Zhang, X.; Zhang, J.; Xu, C.; Jin, Y. Advanced space laser communication technology on cubesats. *ZTE Commun.* **2021**, *18*, 45–54.
8. Araguz, C.; Marí, M.; Bou-Balust, E.; Alarcon, E.; Selva, D. Design guidelines for general-purpose payload-oriented nanosatellite software architectures. *J. Aerosp. Inf. Syst.* **2018**, *15*, 107–119. [CrossRef]
9. Mathason, B.; Albert, M.M.; Engin, D.; Cao, H.; Petrillo, K.G.; Hwang, J.; Le, K.; Puffenberger, K.; Litvinovitch, S.; Storm, M.; et al. CubeSat lasercom optical terminals for near-Earth to deep space communications. In Proceedings of the Free-Space Laser Communications XXXI, San Francisco, CA, USA, 4–6 February 2019; SPIE: Bellingham, WA, USA, 2019; Volume 10910, pp. 24–29.
10. Cheng, X.; Xu, S.; Liu, Y.; Cao, Y.; Xie, H.; Ye, J. Development of an Optoelectronic Integrated Sensor for a MEMS Mirror-Based Active Structured Light System. *Micromachines* **2023**, *14*, 561. [CrossRef] [PubMed]
11. Bagolini, A.; Sitar, A.; Porcelli, L.; Boscardin, M.; Dell'Agnello, S.; Delle Monache, G. High Frequency MEMS Capacitive Mirror for Space Applications. *Micromachines* **2023**, *14*, 158. [CrossRef]
12. Goldman, D.A.; Serra, P.; Kacker, S.; Benney, L.; Vresilovic, D.; Spector, S.J.; Cahoy, K.; Wachs, J.S. MOEMS-based Lens-Assisted Beam Steering for Free-Space Optical Communications. *J. Light. Technol.* **2023**, *41*, 22999592. [CrossRef]

13. Wang, Q.; Ma, J.; Tan, L.; Liu, Y. Influence of the platform random vibration on the beacon in a long-distance laser link. *Opt. Lasers Eng.* **2023**, *162*, 107423. [CrossRef]
14. Wang, X.; Xie, Y.; Liang, H.; Zhong, N. Analysis of Distortion Based on 2D MEMS Micromirror Scanning Projection System. *Micromachines* **2021**, *12*, 818. [CrossRef]
15. Kang, E.; Choi, H.; Hellman, B.; Rodriguez, J.; Smith, B.; Deng, X.; Liu, P.; Lee, T.L.-T.; Evans, E.; Hong, Y.; et al. All-MEMS Lidar Using Hybrid Optical Architecture with Digital Micromirror Devices and a 2D-MEMS Mirror. *Micromachines* **2022**, *13*, 1444. [CrossRef]
16. Cao, Q.; Tan, Y. Online Optimization Method for Nonlinear Model-Predictive Control in Angular Tracking for MEMS Micromirror. *Micromachines* **2022**, *13*, 1867. [CrossRef]

MDPI
St. Alban-Anlage 66
4052 Basel
Switzerland
www.mdpi.com

Micromachines Editorial Office
E-mail: micromachines@mdpi.com
www.mdpi.com/journal/micromachines

www.ingramcontent.com/pod-product-compliance
Lightning Source LLC
Chambersburg PA
CBHW061627210926
51850CB00009B/2464
* 9 7 8 3 7 2 5 8 0 8 4 2 7 *